THE FIRST BOTANICAL COLLECTORS IN NEPAL

The Fern collections
of
Hamilton, Gardner and Wallich

lost herbaria, a lost botanist, lost letters and lost books
somewhat rediscovered

C.R. Fraser-Jenkins

BISHEN SINGH MAHENDRA PAL SINGH
23-A, New Connaught Place
Dehra Dun-248 001 (INDIA)
2006

THE FIRST BOTANICAL COLLECTORS IN NEPAL

ISBN - 81-211-506-4

Citation :

Fraser-Jenkins, C.R. (2005). *The First Botanical Collectors in Nepal - The Fern collections of Hamilton, Gardner and Wallich - lost herbaria, a lost botanist, lost letters and lost books somewhat rediscovered*, Bishen Singh Mahendra Pal Singh, Dehra Dun, INDIA

Address of the author :
Student Guest House, Thamel, P.O. Box 5555, Kathmandu, Nepal
E-mail: chrisopteris@yahoo.co.uk

Printed by Gajendra Singh Gahlot at Shiva Offset Press, and composed by Doon Photo Printers, 14, Old Connaught Place, Dehra Dun, India *for* Bishen Singh Mahendra Pal Singh, Dehra Dun, India.

FOREWORD

The beginnings of Himalayan botany arrived somewhat later than the first flush of Botanical study in India, which began in the South with the arrival of European Botanists in the mid 18th Century, some of whom corresponded with the founder of Botanical and Zoological nomenclature, Karl Linné, or Linnaeus. The Himalayan region was difficult of access, under the control of local powers who were often alarmed into a protectionist policy against foreign intrusions and interests, and the terrain rugged, wild and dangerous to health. This particularly applied to the fiercely independent and forbidden Kingdom of Nepal, a botanical Shangri La waiting to be discovered by Medical Scientists, as by those of other disciplines.

But in the early 19th Century relations, tenuous and difficult at first, began to emerge between the newly consolidated power of the British Raj in India and the Royal Court of Kathmandu, Nepal, recently placed in the premier position by the legendary war-like expansion of the Kings of Gorkha. Against this background various Medical men or Diplomats from British India, who were also the torch-bearers of Botanical Study, managed to reach Nepal adjoined to Political Missions and carried out the first exciting collections revealing its rich and mainly unknown flora.

The first of these was Dr. Francis Buchanan (later Hamilton) in 1803-4, adjoined to Knox's short-lived Diplomatic Mission to Nepal under Rajah Rana Bahdur Shah. Following him came the Hon. Edward Gardner, the first permanent British Resident, or Ambassador in Kathmandu, under one of the conditions of the Treaty of Sugauli, which brought the Gorkha Wars to an end in 1815. Following him was the famous Dr. Nathaniel Wallich, Superintendent of the Hon. East India Company's Botanical Garden at Calcutta, who visited Nepal from 1821-1822. These three Botanists' activities were mainly confined to the Kathmandu Valley due to restrictions placed on them, but they were also able to obtain some interesting collections from the Gossainkund area north of Kathmandu. All of them had an interest in ferns as well as flowering plants and made collections of Pteridophytes in some quantity, which were soon to be much studied as the Victorian penchant for exotic ferny fronds began to take hold back in Britain.

Three important early botanical works were produced that included or dealt with ferns, but unfortunately only one of them was ever published, the other two (by Buchanan and Wallich) remaining in almost total obscurity for nearly 200 years in archives in London and Calcutta, until brought to light at present by the author. The work that was published, in 1824, was by Professor David Don, working on Buchanan and Gardner's collections sent back to London, and was a preliminary study called, *Prodromus Floræ Nepalensis*. Although only a small work, not produced in much detail, it is of great importance in Himalayan and Asian Botany as it provided the first naming and description of part of the Sino-Himalayan flora. Being at an early date, many of the names in it have nomenclatural priority over later names by the mainstream of British and Continental European Botanists who studied the Himalayan flora from the mid-Eighteenth Century onwards. It is hoped that the present work may provide the historical-botanical background to help understand these early Himalayan publications.

Date : January, 2006

C.R. Fraser-Jenkins
Kathmandu

ACKNOWLEDGEMENTS

The author wishes to thank the Trustees and staff of the Natural History Museum, London, for their much appreciated Botanical Research Curatorship in June 2002, which was of enormous assistance in enabling him to continue there for a year of exciting study; he also thanks Alison Paul and Josephine Camus of the Cryptogamic Section (ferns) for their multifaceted help during his stay and at many other times, including a brief visit in April 2004; he also thanks Malcolm Beasley, Head Librarian, and Judith Magee, of the Botany Library, and Ann Datta, formerly of the General Library for their invaluable help in finding even the most elusive of references. His thanks are due, as always to Peter and Margaret Edwards of the erstwhile fern-section at the Royal Botanic Garden, Kew, both for being most willing hosts, and for staying late to help get the impossible tasks done; he also thanks Kate Pickard, of the same institute Archives Section, for her remarkable discovery of merely rumoured ancient tomes.

He is most grateful to Gina Douglas and John Marsden of the Linnean Society, London, for their immediately granted permission and help during visits and in correspondence, and for kindly making available to him Buchanan's *Flora Nepalensis*.

His profound thanks are due to Dr. M. Sanjappa, Director, Botanical Survey of India, Howrah, Calcutta, for opening forbidden doors and for his ongoing interest in saving the remarkable documents so revealed; he thanks Mr. M.K. Deb, Librarian (retd.), B.S.I., for his patience and kindness in helping a sometimes anxiously dedicated pteridologue from the wilds of Nepal; he also thanks Dr. S.K. Basu, of the fern section, B.S.I., Calcutta, for his help while the author was visiting.

He is especially grateful to Dr. Henry Noltie, of the Royal Botanic Garden, Edinburgh, for his knowledgeable discussion of aspects of Indian Botany, and for his interest and subsequent participation in the rediscovery of the Wallich documents at Calcutta.

For their willing advice and indication of historical sources previously unknown to the author, he would like to thank Prof. David Arnold, Dr. David Waterhouse, William Dalrymple, and John Whelpton, all well known authors of British-Indian Historical Research.

He would like to acknowledge the great help he was unstintingly given in previous years while becoming familiar with the complexities of nomenclature, taxonomy, cytotaxonomy and botanical history, at the Natural History Museum, London, as a result of many discussions with the late Mr. Oleg Polunin, the late Mr. Adam Stainton, the late Prof. W.T. Stearn, Dr. Arthur Chater, Dr. Mary Gibby, the late Mr. Jim Crabbe, Mr. Clive Jermy, Dr. Norman Robson, Mr. Eric Groves, Dr. Roy Vickery and Dr. Charlie Jarvis.

A final, slightly guilty thank you goes to his dear wife, Nirmala, for helping the author photocopy no less than forty thousand pages of essential botanical references in the BM in 2002, which has enabled the work to continue, and still more, for patiently putting up with this apparently endless pteridological affliction, instead of departing for the hills ....

CONTENTS

INTRODUCTION

The Nepalese collections of Francis Hamilton (formerly Buchanan), from 1802-1803, the Hon. Edward Gardner, from 1817-1820 and Nathaniel Wallich, from 1820-1821, form the basis of the first botanical publication expressly treating Himalayan plants, David Don's *Prodromus Floræ Nepalensis* (1824). Gardner's part in the work of Don and Wallich has been much overlooked until recently.

Don's important work has caused endless confusion due to the difficulty of locating and identifying type specimens. The same is true of Wallich's plant names, which were mainly validly published by other workers, whose types need to be identified in the various centres they worked from. Summaries of the lives and work of these collectors and authors are given, with the general location of relevant specimens and a Key to recognising Don's types. The pteridophytes of David Don's *Prodromus* have been studied in detail by the author and a full account of them is currently in preparation for publication (Fraser-Jenkins 2006, in prep.).

COLLECTORS IN NEPAL

The first collector was **Dr. Francis Hamilton** (1762-1829), MD, Surgeon, FLS, FRS, Deputy Lieut. of Perthshire, and Chief of the Clan Buchanan, who was known as Francis Buchanan until he succeeded to the Hamilton estates at Leny Castle, near Callander, Perthshire, Scotland, in 1816, through his mother. He changed his name to Hamilton in 1818, by legal proclamation, though his name is often written nowadays (and also sometimes in the 19th Century by him or others) as Francis Buchanan-Hamilton, but incorrectly so as he actually made a straightforward change of surname from Buchanan to Hamilton under the terms of the will. Nevertheless the accepted standard abbreviated author-citation for him given by Pichi Sermolli in Brummitt & Powell (1992) and Pichi Sermolli (1996) is Buch.-Ham. Hamilton (as Buchanan) worked for 20 years surveying and collecting in India, becoming the 4th.

Superintendent, after Lt.-Col. Robert Kyd (1746-1793), Dr. William Roxburgh (1751-1815) and Judge Thomas Henry Colebrooke (1765-1837), of the Botanical Garden at Sibhpur, Howrah, Calcutta, under the East India Company, and after retirement due to ill health, worked on many more important publications from his home in Scotland. The genus *Buchanania* Spreng. was named after him, as were a large number of species of flowering plants and ferns with the epithet *hamiltoniana*.

Francis Hamilton was the son of the physician, Thomas Buchanan, of Spittal, Dumbartonshire, Scotland, and his wife, Elizabeth, daughter of John Hamilton, of Leny. Unfortunately no portrait of Francis is known to exist (Prain 1905) and none was in the family collection at Leny Castle before its demise and sale (Noltie pers. comm. 5 Jan. 2005), though portraits of his two brothers and mother were at Leny, and he was described by his son as being a tall, portly man, with a florid face, white hair and a gammy leg, from an old wound obtained in an action while at sea (Prain 1905, Allen 2002). He married Anne Brooke (or Brock, following Scottish pronunciation) when he was a surprising 59 years old and had one daughter and a son, who later succeeded to Leny. Comprehensive details of his life-history, together with a list of his over 60 papers, have been given by Prain (1905), partly based on Hamilton's (1821, *Trans. Roy. Soc. Edinburgh* **10**: 171 *et seq.*) autobiographical notes. Other works giving details of his life, works and collecting expeditions *etc.* are by Smith (1832), Stephen (1886), King (1899), Britten (1902), Landon (1928), Dawson (1934), Phillimore (1950), van Steenis-Kruseman (1958), Burkill (1953), Hasrat (1970), Hara, Stearn & Williams (1978), Stafleu & Cowan (1979), Mabberley (1977), Desmond (1992, 1994), van Schendel (1992), Noltie (1999), Press & Shrestha (2000), Allen (2002) *etc.* Many of his papers are preserved in the India Office of the British Library, London (Allen 2002). A number of his letters are reproduced in *The unpublished Letters of Dr. Francis Buchanan (afterwards Hamilton) addressed to Dr. William Roxburgh, 1795-1812, from the library of Sir David Prain* (copy at BM!), including detailed descriptions of his Nepalese journey. Others, from 1819-1828, were in the Wallich correspondence sent back from Kew to CAL (as can be seen from the Index to Wallich's correspondence there), some of them having been published by Prain (1905), and might still survive in cupboards in the old library there, though the present author was unable to find them there in Dec. 2004. A full set of Hamilton's original Nepalese and other letters, giving details of his visit there, are in the Smith correspondence at LINN, and other letters are in the Scottish Rccords Office (Desmond 1994). In addition to those mentioned here he wrote many further important detailed accounts of his surveys, which survive today, often unpublished.

Like many early botanists in India, Hamilton was a medical surgeon. He did his MD at Edinburgh, under Professor John Hope, publishing a paper on Malaria, during which time he became a friend of Sir James Smith of Norfolk and London. At first he made two or more sea-voyages to the Comoro Islands, India (Bombay), the Philippines and Celebes *etc.*, as a medic, starting in 1785, but then in 1794 went out again to join the Bengal Medical Service of the East India Company.

In India he became a highly accomplished surveyor and excellent naturalist, draftsman, making many major collecting expeditions throughout the region for the next 20 years. His interests were extraordinarily wide by today's standards, not only in the natural sciences of botany, zoology and geology, but also in anthropology, archaeology, religion, history, linguistics, medicine, economics, ethnology and genealogy, and mapping and surveying, as can be seen from the list of his publications given by Prain. Yet his knowledge and work were not only of such surprisingly wide scope, but also of outstanding depth in all these subjects and present a detailed excellence, seldom matched in his day. He published many reports of his regional surveys (including detailed descriptions of the now famous Buddhist site of Bodhgaya, in Bihar) and later in retirement in Scotland a number of important botanical works, as well as his superbly illustrated and painted work on Indian fish, *An account of the fishes found in the River Ganges and its branches* (1822); his major botanical works were his slightly incomplete *A Commentary on the Hortus Malabaricus* (1822-37), and *Commentary on the Herbarium Amboinense* (1823-1831), though he did not identify van Reede's and Rumphius' ferns in them. In addition he left a number of detailed unpublished manuscripts and original paintings, now preserved in the India Office Library, London, and more so at the Linnean Society of London, given by him to Sir J.E. Smith. However it must be said that Hamilton "really was most unlucky in his choice of collaborators" (H.J. Noltie pers. comm. 24 Feb. 2005) as most of his manuscript-names and much of his scientific work were frequently sat on and ignored by those he sent it to. Other parts of it were taken over by his contemporaries, Sir Joseph Banks, Sir James Smith, Prof. David Don and Dr. Wallich, sometimes without proper attribution or with altered names. A considerable portion of his work has remained unknown to the present day, though still preserved and coming to light in recent times. Despite his many successful publications, several other manuscripts and fine illustrations of his still remain unpublished and unavailable to most botanists.

Hamilton began his travels for his employers, the East India Company by being sent to the erstwhile Kingdom of Ava, Myanma (Burma), in 1795, with a diplomatic Mission under Capt. Michael Symes (Symes 1800) of the 76th Foot Regt., travelling via the Andaman Islands. His herbarium-specimens from this expedition were sent to Banks, the President of the Royal Society, and are today in the Banks herbarium at the Natural History Museum, London (BM). His unpublished *Enumeratio Plantarum Quas in adeunde civitatem Barmanorum regiam, et dehinc redeundo Anno MDCCXCV observavit Franciscus Buchanan MD* (1797) (copy at BM! and see Britten 1902), gives details of his plant-collections there, and several were also illustrated by him in Symes' work. Hamilton's account detailed some 10 species of ferns, though these were subsequently overlooked by later authors. In 1798 he made an expedition to what is now S.E. Bangladesh, in the Chittagong Hills, travelling first around what is now Cox's Bazaar, then to Sitakunda ("Seetacoon") Hill, north of Chittagong (where Roxburgh had collected before him and which Sir J. Hooker visited later), then proceeding up the Karnaphuli ("Karnafouli") river, east to Rangunia, near modern day Rangamati and then further up to Barkol, before returning and heading north to Comilla in what is now central Bangladesh, but was then in "Tiperah" (Tripura). The herbarium-specimens from this expedition were again sent to Banks and are in the Banks herbarium at the BM. His long-unpublished account of the journey was entitled *An Account of a Journey undertaken by Order of the Board of Trade through the Provinces of Chittagong and Tiperah, in order to look out for the places most proper for the cultivation of spices, by Francis Buchanan MD* (Buchanan 1798, *ined.*), and has recently been published and expounded in excellent detail by van Schendel (1992), including a number of botanical details. But his herbarium-material has not been investigated beyond listing it in the data-base of the Banks herbarium (now on the web) by the Natural History Museum, London (BM).

His next visit was to the Nepalese border in 1800, with a diplomatic mission intending to go on up to Nepal, but was cancelled for political reasons before they could do so. He was sent instead to Mysore and Malabar in South India, with the young Duke of Wellington (as he was later to become), from 1800-1801, to draw up a three-volume report of the flora, economics and geology of the area that had just been wrested in 1800 from Tipu Sultan of Mysore at the Battle of Seringapatam by the East India Company (see Noltie's (1999) remarkable note concerning the rescue of one of Hamilton's books there), entitled *A Journey from Madras through the Countries of Mysore, Canara and Malabar performed under the orders of the Most Noble the Marquis of Wellesley etc.* (Buchanan 1807). His botanical-collection from there was brought back to London by him when on leave in 1805, and given to Sir J.E. Smith, President of the Linnean Society, so should now be mainly at LINN, though, if collected, no ferns from this survey appear to be there nowadays (Savage 1963, Catalogue, *ined.*, seen by the author in 2003). Smith published about 12 of the flowering plants in his *Exotic Botany* (Smith 1804-1808, see his note in vol. **2**: 73, *t.* 97) and a few others in Rees' *Cyclopaedia etc.* (Smith 1814, 1819 *etc.*), but the rest of the collection was left unknown. The status of the Nepal specimens in LINN as isotypes of Don's species has been mentioned by Stearn (1960: 180, 1988:205).

The year after his south Indian visit, Hamilton became the first botanical collector to visit Nepal, from March 1802 to March 1803, having been sent to join the third diplomatic mission to Nepal (following Kinloch's in 1767 and Kirkpatrick's in 1793 (Kirkpatrick 1811)), at the instigation of Richard, Marquess Wellesley (the Governor General of Bengal), and Lord Lake, the Commander-in-Chief. This was led by Capt. (later Major) William H.D. Knox, of the Bengal Cavalry (Prain 1905,

Landon 1928, Phillimore 1950, Hasrat 1970), who had been appointed "Resident at the Court of Nepaul" [*sic*], along with the surveyor, Capt. (later Colonel) Charles Crawford, commanding the Resident's Escort, and Mr. Blake, Geologist. Hamilton was assisted in his collections by a most accomplished Bengali Brahman from Calcutta, Babu Ramajai Bhattacharji, who later accompanied him on other surveys. However, the British presence in Nepal, which had been at the invitation of Rajah Rana Bahadur Shah while temporarily in exile in Varanasi (Benares), was highly unpopular with the Nepalese, who were strongly suspicious of the great power developing adjacent to them. Hamilton and the party were thus largely confined to the British Residency near Narayanhetti ("Narainhatty"), Kathmandu, throughout their eleven month's stay in Kathmandu (fourteen months in Nepal altogether), with only occasional excursions in the Valley or nearby. However he was able to collect on the way up to Kathmandu, their route being detailed in his letters to Roxburgh, though it was less easy for him to collect on the way back as they had to leave in haste for their own safety, if much to Hamilton's strongly expressed relief to be out of the tiresome restrictions in Nepal. He was also able to make one excursion outside the Valley, to Nuwakot ("Noakote"), near Trisuli Bazaar, to the north-west of the Valley, on the route north towards Gossainkund (and today's Langtang), but collected rather few plants and no ferns there.

The party began from Bankipore, near Patna, on 20 Jan. 1802 and went north via Tirhut to Dacca, near the Nepalese border, by 28 Jan. 1802. In early February they waited at Kuchurun, near Knox's camp at Ghossaini, "the last place in the Company's territory" in India, to be accompanied up to the Valley by some "Nobles of Catmandu", including the Rajah of Betoul (Butwal), Damal Damodar Pande, Bum Shah Chautaria and Tribhuvan Singh Bisnait (Phillimore 1950: 70, Hasrat 1970). Pande and Chautaria had led the disastrous retreat of the Gorkha army from Tibet over the Hatia Pass in 1790, and Pande later became the first ruling Prime Minister of Nepal (his son being instrumental in the unfortunate destruction of the great Prime Minister, Bim Sen Thapa, in 1839). The Nepalese party came down to the Indian border to guide and observe the party, which began the march into Nepal on 11 Feb. 1802 via Ghorassan. Hamilton had hoped that Roxburgh's son, William, would join him as an assistant, but when they halted at Bassaria (2-11 March 1802) Hamilton wrote "Capt. Knox informed me that very serious objections have been made by the Nepal Govt. to the number of Europeans now in company with him and that the strongest objection would be made to the coming of any more, who had no evident employment but that of spying the nakedness of the land", so William Roxburgh was not able to join in the expedition. However the party were allowed to continue via Baroni (11 March 1802), Norcotera (26 March 1802, slightly back into India), Danawara (27 March 1802), Ghor Porsera (29 March 1802), Ettounda (Hetauda) (2-4 April 1802), Chitlong (11 April 1802), and Tancote (Thankot, the present entrance to the Valley) (15 April 1802) ("we arrived here yesterday and are in sight of Catmandu in a large, bare, ugly valley resembling many of those in Scotland before the introduction of fences or other improvements" - sentiments modern visitors would find themselves much in opposition to, as the Valley opens up spectacularly before them at Nagdunga, Thankot!). They then reached Catmandu (Kathmandu) (17 March 1803), where Hamilton listed all the plants he had collected en route in April, including some unnamed ferns.

While in Kathmandu, though under close observation from the Nepalese authorities, he was able to visit and collect in a few localities around the Kathmandu Valley during the rest of that year and early in 1803 and proceeded to send letters to Roxburgh in Calcutta over many months detailing the plants *etc.* he was able to discover in the Valley. As interesting examples, in his letter of 23 July 1802 he sent "Also the roots of a fern, the young shoots of which are eaten by the natives. I believe it does not differ from the common fern of Europe or *Pteris aquilina.*" [but from the date and local knowledge it must be guessed that this would actually have been *Diplazium esculentum* (Retz.) Sw., since, unlike that, bracken proper (*Pteridium revolutum* (Bl.) Nakai, in Nepal *etc.*) is not eaten in Nepal and is seriously toxic]. In his letter of 2 Dec. 1802 from Kathmandu, his collection number 9/78 was of "Plants of *Polypodium tuberosum* B[uch.-Ham., *ined.*] [= *Nephrolepis cordifolia* (L.) C.Presl, *nom. cons.*] of which the Newars are fond in the warm season. They eat these bulbs raw." After what Hamilton described as a tiring and much restricted stay, which he was happy to see to a close, his party left Kathmandu rather precipitously on 18 March 1803 and proceeded south via Pharping ("Pherphing"), near Dakshin Kali temple, and Kangoo. Hamilton's letter from Laalgunga Singhea, 3 April 1803, read: "As was said, we all marched from Catmandu on the 18th March and on the 28th Capt. Knox arrived at Suggouly (Sugauli, near Roxaul, in Bihar). Here I was relieved and set out next day for Patna and have now reached within two days march from that place".

It is fortunate and typical of his careful work that Hamilton took the trouble to localise his collections in sufficient detail, in contrast to the very large collections of Wallich which only mention Nepal ("Napalia") as their locality. Most of Hamilton's localities can be readily identified nowadays, and most are also mentioned in his important economic and historical account of Nepal, *An Account of the Kingdom of Nepal and of the Territories annexed to this Dominion by the House of Gorkha* (Hamilton 1819), that he wrote after his retirement to Scotland. Madden (1856) later described some of the plants mentioned in the Account. The Pteridophyte localities written on Hamilton's specimens were Suembu Nepalensium (1st and 15th May 1803), or Swayembhunath hill, on top of which sits the famous "Monkey Temple", on the western edge of Kathmandu, many of his species still occurring there today (*Aspidium mucronatum*, *A. cornucervi*, *Nephrodium sparsum* and *Pteris cretica* were so reported from this collection by Don (1824)); Lahuri near Chitlong (11 April 1802), on the old route directly northwards over to Kathmandu from the *terai* plains via Hetauda (*Polypodium lineare*, *Aspidium discretum* and *Asplenium laserpitiifolium* were so reported by Don); Hethaura Nepalensium (3 April 1802), the large town of Hetauda, in the Nepal terai at the foot of the hills due south of Kathmandu (*Nephrodium cochleatum* was so reported by Don); Nilkantha inter alpes Nepaliae, now Bouddhanilkantha and its famous temple, at the north-western edge of the Valley (*Hymenophyllum ramosissimum* was so reported by Don); and Kharbu vicum [village] Nepaliae, now ?Kharpu, near Dhulikhel, or ?Khargu near Chitlong (*Woodwardia radicans* was so reported by Don). The modern identity of these fern-names is given by Fraser-Jenkins (in prep.).

It may be noted that in many cases the locality Narainhetty Nepalensium is given, as commented on in Don's preface. Narayanhetti is the proper name for the current location of the Maharajah, or King of Nepal's main Royal Palace (built in 1847 by Maharaja Jang Bahadur, the ruling Rana Prime Minister, and later made into a Royal Palace) and its surroundings, also known locally, nowadays, as Ratt Durbar (meaning "red palace"), just east of today's Thamel (formerly Tamabhil) area. Some of Hamilton's specimens are dated as from Narainhetty on 28 Dec. 1802, 23 Jan. 1803 and 24 and 25 Feb. 1803, but it seems perhaps rather unlikely that all of the species reported from there really occurred there (*Polypodium flocculosum*, *Aspidium squarrosum*, *Asplenium normale*, *A. falcatum*, *Diplazium falcatum*, *Hemionitis pothifolia*, *Lomaria decomposita*, *Leptostegia lucida*, *Vittaria revoluta*, *Cheilanthes dealbata*, *Adiantum lunulatum*, *Lycopodium circinale*, *L. setaceum*, *L. obtusifolium* and *L. serratum* were so reported by Don), even allowing for its being a settlement outside Kathmandu and the adjacent ridge being naturally afforested at that time. In fact Narayanhetti was the location (as given at that time, before the name Lainchaur came into existence) of the British Residency in Kathmandu, where Hamilton stayed, calling it Ranikabag (Queen's Garden). The site had been given to the British party for the temporary British Residency of Knox and his party, later becoming the permanent Residence of today (illustrated from a distance by Hamilton (1819) and the building as it was in 1833 by Waterhouse (2004)), and the present author is not absolutely convinced that all the plants so labelled were really collected from that locality, rather than it being written by Hamilton more in the way of an address where he was working from. The present day Narayanhetti palace was not at that time the Royal residence (which was the palace at Durbar Square in the centre of Kathmandu), as it is nowadays, but is adjacent to where the British Residence became situated and the whole area was then known by that name. The Embassy land was given by the Nepalese as an out of the way, fever-ridden (at that time only), uninhabited, forested ridge, even reputed to be haunted by ghosts, some miles outside the Kathmandu of that time. The only other building of substance then at Narayanhetti was the abode of the dispossessed minor Royal relations, the Chautarias (who were mainly exterminated in the Kot massacre of 1846), which much later became the forerunner of the modern Royal Palace. The area of the Embassy land is about a hundred metres, at its nearest corner, from the grounds of Narayanhetti Palace, although it is now in the current area of Lainchaur, while Narayanhetti today refers only to the Royal Palace.

Following his Nepal expedition, Hamilton returned to Calcutta in 1803 to work on his collections and manuscripts, with animals often drawn from the Marquess Wellesley's menagerie at Barrackpore, just north of Calcutta, which later became the origin of Alipore Zoo (Phillimore 1950: 384) and was largely created by Hamilton. The zoological drawings he made there should survive in the Asiatic Society of Calcutta. His Nepalese and south Indian collections were then brought back to London with him in 1805 when he returned for two years of leave, accompanying the Marquess Wellesley who was retiring from his post as Governor General of Bengal. He was able to work on them there and presented to Sir James Smith, President of the Linnean Society, a fine and detailed, if preliminary *Flora Nepalensis* (Buchanan 1802-1803), written while he was held up in Kathmandu. This important work gives full details and names to his collections, in more detail than Don's work, including ferns (among 307 species of plants treated (Mabberley 1977)), and is listed in LINN, where it is preserved today, as "Ms 401 *Flora Nepalensis*, incomplete Flora of Nepal, commencing with Cryptogams and terminating with *Lysemachiae*, made in 1802-03, based on collections taken chiefly near Kathmandu. 168 pp. with loosely inserted pencil sketch of a fungus and an Ms note. With a typescript note on the flora by A.H.G. Alston 1945 (this flora was presented to J.E. Smith in 1805 with Nepal herbarium of 1500 plants)". Ms 403 includes drawings and paintings of birds and animals, and also 182 sheets of watercolours of *Plantarum Nepalensium Icones Pictae* (with drawings and paintings of several of his ferns from there). Hamilton's herbarium-collections from Nepal were given partly to the private herbarium of Sir J. Smith, at the Linnean Society. Hamilton gave him some 1500 herbarium-specimens (Smith 1805, Prain 1905, Burkill 1953), which Press & Shrestha (2000) and some other modern sources say consisted of his material from both S. India and Nepal, though Smith (1805) himself and Prain (1905) said were from Nepal only; Hara, Stearn & Williams (1978) stated that Buchanan collected 433 specimens from Nepal, thus if only these were given to Smith, the rest of the 1500 would have been from S. India. Apart from odd gifts from Smith these should now be in LINN, but only 192 sheets, less than half, survive today in LINN (Press & Shrestha 2000). However Hamilton gave another set to Aylmer Bourke Lambert, which is now in BM; this set is evidently a more complete one at the present time and of the 363 sheets (31 being Pteridophytes) ticked in Don's book as purchased for the BM, Press & Shrestha were able to find 285 sheets of flowering plants, while the present author has found 32 sheets of Pteridophytes. Apart from some occasional gifts by Don from Lambert's herbarium (*e.g.* now at OXF), no material from this collection exists elsewhere.

Although several reliable sources reveal that the main set was given to Smith, it is curiously evident that there are no ferns in LINN from Hamilton's Nepal collection today, if they were ever there. How this can be the case has never been explained or commented on and appears to be a curious anomaly, disagreeing with the earlier literature-sources. Burkill (1953), amplifying Prain's (1905) comments to the same effect, pointed out that Smith did virtually nothing with the Hamilton collections, and indeed it can be seen with hindsight that their being given to Smith had unfortunate results. Now that the surviving Hamilton collection at LINN have been assessed (Press & Shrestha 2000, Fraser-Jenkins in prep.) it appears that either large portions of it must long ago have been lost by Smith, or that Hamilton gave the fern part of his collection exclusively to Lambert. Fortunately the set given to Lambert still contains nearly all of the fern specimens, mostly with full details of locality written by Hamilton and often with their dates and was the important set utilised by Don (1824) (see below, *sub* Lambert). In fact Don himself stated in the introduction to his book that of Hamilton's collections the greatest part existed in Lambert's herbarium. The Lambert set, now the main one, is still preserved today in the BM. The situation is thus quite the opposite way around from Prain's saying that even after searching for 12 years he had been unable to locate the Lambert set, perhaps then still lying unavailable in unincorporated boxes at the BM, and erroneously deducing that it must have been destroyed (Prain 1905). Concerning the Smith set, which he described as the main set (as it must once have been), but, significantly, did not himself see, Prain said "the fine original collection given to Smith should still be in Smith's herbarium; if not there, the authorities in charge of the Linnean Society's collection should be able to say where it now is". But today it remains unexplained as to why there are considerably fewer specimens at LINN than in BM, and why there are no pteridophytes at LINN. Hamilton's ferns were not catalogued there by Savage (1963, *ined.*), nor have they turned up anywhere else. In the Data Base made at LINN from Savage's vol. **6** (pteridophytes) there is not a single Hamilton specimen from Nepal, the only Hamilton fern being one from India in 1800. In fact the only Nepalese fern-collections at all there are 9 specimens from Wallich, dated 1819, and thus collected by E. Gardner, and two later Wallich sheets collected in 1825. Thus at least the Nepalese fern-parts of Smith's herbarium, if he was actually given ferns, appear to be a second "lost herbarium", but unlike Lambert's, have not been refound.

While pointing out the difficulty of identifying Don's names as being a major problem in Indian (and Chinese) pteridology, the late renowned Prof. R.-C. Ching (1984), of Beijing, stated that his types may have been destroyed by burning, without giving further details ("It is said that part of Don's type specimens was latter [*sic*] lost in a fire and what has since been left at the British Museum (Natural History) are rather fragementary [*sic*]. Through the effort of Mr. C.R. Fraser-Jenkins, I saw last year a few Don's types specimens and, to my great surprise, they turned out to be distinct species (see Appendix [where four new combinations or names were made]). There are still a number of Don's species remain *species dubia* pending further searching at his herbarium in British Museum (Natural History). Unless their identity is established in the future, the fern flora of the Himalayas will nomenclaturally remain in confusion as it is now."). Prof. Ching-*ha* had previously mentioned this as a question seeking confirmation from the present author, who had been explaining various Don species to him in Beijing in April 1980, but further discussion with him revealed that he knew no further details, though he thought he might have been told this, perhaps by Christensen, during his visit to European herbaria in 1932. Ching had felt that it could be the explanation as to why he had not seen (actually had not been able to recognise) Don's types during that visit. However this must have been based on Prain's (1905) mistaken idea that the Lambert sets used by Don were destroyed as unsellable by Pamplin and perhaps burnt. The real explanation for previous workers' lack of knowledge of the types is that it came about from the difficulty of recognising them, enhanced by Don's not having annotated any but a small handful of the specimens he saw. On knowing which dated material to look for and what the name was taken to be by other authorities, it is nearly always possible to find the relevant original (fern-) material all safely preserved at the BM. Only in a few pteridological cases has further search beyond the BM been necessary.

When Hamilton returned to India in 1807 he embarked on a detailed and exhaustive survey of Bengal (Hamilton 1833), including the area now in Bihar and eastern Uttar Pradesh, adjoining the Nepalese border. This major work, involving great distances of travel, with a team of collectors, artists and surveyors, took him seven years to complete. Places he visited were listed by Burkill (1953: 865) and included the whole mid and lower (deltaic) Gangetic region, Eastern Uttar Pradesh, Bihar, north Bengal, Assam and Meghalaya. He also sent some collectors into Nepal (Morang Hills *etc.*), while stationed near the border, to obtain a few higher altitude Himalayan species. The results of the survey were later republished as *The History, Antiquities, Topography and Statistics of Eastern India etc.* (Martin 1838); *An Account of Asam with some notices concerning the neighbouring territories* (Hamilton 1820); *Some notices concerning the Plants of various parts of India, and concerning the Sanscrita names of those Regions* (Hamilton 1826); *Journal of Francis Buchanan (afterwards Hamilton) kept during the Survey of the Districts of Patna and Gaya in 1811-1812* (Jackson 1925) and various other papers. The botanical specimens from his Bengal collections were sent from Calcutta to East India House in 1822 (Mabberley 1977) and were then split into two main sets by Wallich on his long leave to Britain as from 1828. The first set was permitted by the E.I.C. to be included in Wallich's Numerical List (1828-1849) and distributed by Wallich from 1828-1832. These are now mainly in K-W, K, BM, G, BR, P-JU, AWH (among other places) and were in a good set of Wallich specimens brought by Thomson to CAL in 1855, now mostly destroyed. Hamilton's second set of the Bengal collections, along with a copy of the catalogue, was given by him to Edinburgh University, and is now at E. Paintings from the survey based on Hamilton's drawings and painted by Indian artists are partly at Kew, with a few at Edinburgh. The whereabouts of the Hamilton drawings was later the subject of mistaken personal criticism by M'Clelland and Griffith of Wallich's having brought them to Britain; but today we can only be thankful that Wallich thus effectively rescued most of the material at Calcutta and saved this important part of Indian cultural and scientific heritage, which would otherwise most likely have been destroyed by neglect and the climate (though a set of Roxburgh paintings survives there, but little else). One of these paintings was of *Ceterach indivisum* Buch.-Ham., *ined.*, syn. *Grammitis hamiltoniana* Wall., *nom. nud.*, now called *Colysis pedunculata* (Hook. & Grev.) Ching (Nooteboom's 1997 teatment of this species as *Leptochilus macrophyllus* (Bl.) Noot. is not accepted here), and was painted by the justly famous Indian artist Vishnu Prasad, from a Hamilton specimen from Kamrup, Assam, collected in *c.* 1809; it has been finely illustrated by Noltie (1999). The original catalogue of the Bengal collection was brought to East India House by Wallich, who made a copy for his own use, which was later returned to Calcutta. Wallich was also allowed by the East India Company to loan it for copying to the Linnean Society, who were not only given a numbered set from Wallich's distributions (now in K-W), but also in 1833, the large remnants of specimens from the Indian Museum, East India House, Leadenhall Street, London. The copy of the catalogue at LINN is entitled, *Catalogue of dried plants collected, presented to the Museum of the Hnble. East India Company, and arranged according to the system of Linnaeus, By Francis Buchanan MD* (copies at LINN! CAL, as confirmed from their old library catalogue, discovered by the present author in 2004, but perhaps now destroyed there, and E). It contains 91 species of pteridophytes, including many of Wallich's Mauritian collections of 1812, in 14 pages, but the fern-part is actually of less botanical interest than his other work, mostly containing more common species under well known names, which are often misapplied.

In 1814 Hamilton was appointed Superintendent of the East India Company's Calcutta Botanic Garden, succeeding his friend, Dr. William Roxburgh. But he was already bedogged by serious ill health, like many British officers before and since who had the much dreaded misfortune to be posted in Calcutta - many others of whom merely ended up in the South Park Street Cemetery with their lives cut short by Malaria, Cholera or Dysentery. He was therefore allowed by the E.I.C. to retire early to his home in Scotland the next year and was replaced at Calcutta Botanical Garden by Wallich. After gradually recovering his health at home by 1816, Hamilton continued selected parts of his work. However, he had become considerably disgruntled by the East India Company's refusal to allow him to bring back and complete work on his paintings of fish and other materials, as well as by the failure by Smith and others to make proper use of his collections. As a result, on his return to Britain he decided to give up his work and hand over the remaining specimens he had with him to East India House, in 1815, where they joined his previous collections (later distributed by Wallich or passed to the Linnean Society) and he did no further work on them for a number of years. But by 1820 he was able to obtain access to his Bengal specimens there and worked again on the completion of some of his manuscripts, as well as taking up his major new work on the interpretation of Rumphius' (1741-1755) *Herbarium Amboinense* and van Rheede's (1678-1693) *Hortus Indicus Malabaricus*, both of which commentaries were later largely published. But he no longer worked on his Nepalese collections. In October 1821 he still felt unable to attempt to work further on them when invited to do so by Wallich (in March 1821) (Prain 1905), as the material was "in a sort lost as having been given to Sir J.E. Smith who is rather indolent and not likely to publish any considerable part of what he has" and he had "not a single note respecting any of the plants I brought with me from Nepal - Smith has the whole". He therefore preferred to work on his Commentaries on *Hortus Malabaricus* and *Herbarium Amboinense*. The subsequent publication in 1824 by Don, whom he had initially encouraged with some enthusiasm, of many of his species with changed names and some others attributed to Smith instead, was a further, final blow.

When Hamilton died at Leny Castle in 1829, a year after Smith, he left behind him an enormous scientific, geographical, economic, historical and artistic legacy. The publication of much of it must have given him considerable satisfaction, but one of the most interesting aspects remains his important unpublished material, especially his fine drawings and paintings and his virtually unknown *Flora Nepalensis*. The author is currently engaged in studying the ferns of this work in connection with his ongoing study typifying David Don's ferns (Fraser-Jenkins, in prep.).

Aylmer Bourke Lambert (1761-1842) FRS, FLS (Lambert (1828), Schultes (1830), where he is rather charmingly called "Count Lambert", Sotheby (1842), Lindley (1842), Thiselton-Dyer (1891), Lee (1892), Renkema & Ardagh (1930), Miller (1970), Stafleu & Cowan (1979); Lambert letters at Kew (Thiselton-Dyer 1891: 326) and some in CAL!; portraits at LINN!, Miller (1970), Stafleu (1972)), mentioned above, was a wealthy patron of Botanical Science, who built up an extensive and most important private herbarium of about 50,000 specimens, one the finest in Britain at the time, and lived at his hereditary family seat and garden, Boyton House, Heytesbury, Wiltshire, with his fine library, herbarium and museum at his town-house at 26, Lower Grosvenor Street, London. Among many famous original collections from around the world, the herbarium, which was open for visitors, held a number of collections sent to him from the Indian subcontinent. Most importantly, these included the material published by Don (1824), namely the second and only other set of Hamilton's Nepalese collections (along with some specimens of his from S. India *etc.*) and also a set of the early Wallich Nepal specimens collected by Gardner. Yet despite its importance his herbarium became known in more modern times as a famous "lost herbarium", due to its being split up and sold at auction after his death; it is only since the detailed exposition of Miller (1970) that the history and contents of the herbarium have again come to light along with the whereabouts of the surviving sets of specimens.

The precise contents of his herbarium can be seen from Don's detailed list of it (Don 1828) in Lambert's *Description of the Genus Pinus*, where he also described how they were "glued on a single half sheet of stout folio writing paper" and arranged according to the Linnean system and with a General Index. It included about 2000 species (*i.e.* specimens) from Nepal, "the greater part of which are entirely new," which he listed as to their families or genera with the number of species already determined, mainly by him. The Nepalese pteridophytes (*i.e* both Hamilton's and Wallich's collections) were estimated to consist of 100 species (specimens) of *Filices* and 7 *Lycopodineae*, and Don (1824) actually described 87 Nepalese pteridophyte species. Collections from India included numerous specimens from T. Henry Colebrooke (Chairman of the Asiatic Society of Bengal and Superintendent of Calcutta Botanic Garden) and William Roxburgh (Superintendent of the Calcutta Botanic Garden), while the collections from Nepal were listed as:

"29. Dr. Francis Hamilton (formerly Buchanan), so justly celebrated as a traveller and naturalist, liberally presented part of the fine collection of specimens made during his residence in that highly interesting country, Nepal, in 1802-3; with many others found by him in Mysore, Cannara, Malabar *&c.*"

"30. Dr. Wallich, the indefatigable superintendent of the Botanic Gardens at Calcutta, has enriched the Herbarium with many valuable collections from Nepal and various parts of India, as well as from the Calcutta Garden."

Lambert published many botanical and general works, of which the most famous was his sumptuously produced and painted *Description of the Genus Pinus* (Lambert 1803, 1824, 1828, 1832, 1837, 1842), of which the 1828-37 edition., vol. **2** (1828), contains the important description by Don of the contents of the Lambert herbarium. He was a founder-member and later Vice-President of the Linnean Society and was held in great respect by all the scientific world of his day. The genera *Aylmeria* Mart. and *Lambertia* Smith were named after him, as was *Pinus lambertiana*, among other specific epithets, and the "Lambert Cypress". It is a sad fact that his determined dedication to Botany, even in the face of a decline in his income from the Jamaican sugar-trade, eventually ruined him financially and used up all the Capital from his mother's rich estates in Ireland and Jamaica, so that at his death his collections and library had to be sold to meet the debts instead of being given to the BM, as Lambert had intended as a first option.

The sale of the Lambert herbarium, library and other contents, as well as the town-house itself, was conducted by the premier London Auction House, Sotheby's, over three days in summer 1842, and was attended by many British botanists as well as agents for foreign botanists (see Hooker 1842). One can only whistfully imagine being seated in

Sotheby's improvised sale-room in Lower Grosvenor Street that Monday afternoon, June 27th to Wednesday June 29th in 1842, and once more being able to seize the chance to raise a hand to bid for those vitally important specimens, at the princely sum of between 14 shillings and a Guinea per lot of up to 6 or 7 bundles! Apart from the house (at £2000), the sale made £1170 and 4 shillings, though of course this would have been a large amount by today's standards and such a sale today would fetch millions. The catalogues of sale have been preserved (Sotheby 1842) and Robert Brown, of the Natural History Museum, London, annotated his copy at BM (!), as did Hooker at K (!). Rough details of each lot and the names of purchasers were thus preserved and Miller (1970) was able to make an outstanding in-depth study of their subsequent history. Brown himself bought the Hamilton herbarium for the Natural History Museum (Lot no. 286, described as "*A large Collection of Plants from Nepaul, Mysore, and Malabar, by Hamilton, with the Cabinet in which they are contained, about 500 species. This Herbarium of Hamilton's supplied the materials for Prof. D. Don's Flora of Nepaul*"); it cost £9 and their purchase fortunately preserved the most important set of Don's types at the BM. Strictly speaking, these specimens should be selected as lectotypes instead of holotypes, since there can be other specimens in existence in LINN, OXF, FI-W, P-JU *etc*. In the BM copy of Don's book the former Keeper of Botany, James Britten, has marked with a tick those specimens he could readily identify from the sale, and J.J. Bennett annotated the BM copy of the catalogue that 340 of the total specimens purchased were noticed in *Prodromus Floræ Nepalensis*, with an additional 105 Nepalese plants and 127 Indian plants apparently not listed there. But some specimens without names could not be identified by him, as also the early Wallich set, so are not ticked despite being in the herbarium.

The other half of the material used by Don was the early Wallich set of Lambert's, actually collected in Nepal by Gardner and his team (see below) between 1817 and 1819, and sent from Calcutta in 1818 and 1819, prior to Wallich's own, numbered collections of 1820-1821. This was a rather larger collection than Hamilton's, though frequently without the name of the species (written on later, often in pencil and by more than one hand) and only bearing the locality "Napalia" and name

Wallich, but with the date, from 1817-1819, which is very important in enabling them to be distinguished them from Wallich's later numbered set. The early Wallich material from Lambert's herbarium was not ticked into the BM copy of Don's Flora because it was difficult to identify which species the specimens belonged to due to not having names on them and their being confusable with Wallich's later main sets. But the material was indeed acquired by the BM and was definitely that used by Don; when the whole set of these ferns is studied in BM it is usually possible from Don's descriptions, the pencilled names and by a process of elimination to recognise which name each specimen belonged to. Yet because it was difficult to recognise them various authors have been unable to identify Don's types based on Wallichian material, or mistakenly took Wallich's later numbered material from his *Numerical List* as being his types. Fortunately Brown also purchased this Wallich material from the Lambert Sale, in Lot nos. 130 ("A large bundle of East Indian (chiefly Nepaul) Plants"); 131 (Ditto ditto, Wallich's Plants"); and 258 ("Wallich's Plants, about 750 species, as arranged and mounted by Mr. Lambert, with large cabinet, and a copy of Wallich's Catalogue of East Indian Plants"). It was from one or more of these three lots that most of the original early Wallich material came to the BM, while other parts of the three lots may have been from Wallich's later, numbered sets. It is likely that the only other Nepalese material of Wallich's in the Sale was in Lot. nos. 97 and 99, which were smaller lots containing 10 bundles of "Wallich's Plants of Nepaul, *&c.*", and that the other bundles mentioned by Miller as being bought by Lemann and Pamplin, instead of Brown, were actually Indian plants proper, as stated in the Sale Catalogue. In fact nearly all Don's pteridophytes based on Wallichian collections are represented by specimens of the correct date in BM (Fraser-Jenkins in prep.). C.M. Lemann's herbarium went to Cambridge (CGE) and consisted of Wallich collections (Gilmour & Tutin 1933), though the author has not yet examined them to ascertain their provenance and date. Lots 97 and 99 were bought by William Pamplin, a Soho (45, Frith Street) bookseller, who became editor of *The Phytologist* and in 1863 retired to start an unsuccessful Welsh Botanic Garden at Pen-yr-Llan, near Llandderfel, Caernarvonshire, N. Wales (Anon. 1899, Stafleu & Cowan 1983). It was these that Prain was referring to when he erroneously stated that the Lambert herbarium had been mostly bought up by Pamplin and thence destroyed when he found he could not sell them. Miller found a letter of Pamplin's from 1894 (when he was 88 years old, he died in 1899) in the BM where he said that he was commissioned to act as an agent to several botanists, but could no longer remember who; but Miller also found out that two of the commissions were for H.B. Fielding, of Stodday Lodge, near Lancaster, England (specimens now in OXF), and J.F. Klotzsch, of Berlin Botanical Garden (specimens now in B, where the pteridophytes survived the massive British World War II bombings in 1943). Steinberg (1977) has documented the further acquisition of Lambert herbarium material from the sale by P.B. Webb, residing in Paris, which is now preserved in his herbarium in Firenze (FI-W), this may have also have been bought through Pamplin, but Miller (1970) was unable to catalogue any material relevant to Nepal through the sale. Steinberg found no Hamilton material in FI-W, but listed material from Nepal collected by Gardner and dated 1818 in the Webb herbarium, as well as later Wallich specimens. He also mentioned that during his lifetime Lambert had sent duplicate specimens of some of his material to Webb (Steinberg 1977: 8) and that this included Wallichian Nepal specimens (Steinberg 1977: 41), which must be the origin of Gardner's material in Firenze, which can be cited as isotypes or isolectotypes of Don's names. Stafleu & Cowan (1988) list that early Wallich material is also in Prague (PR), but this may be in error for BR, where there is early Wallich material obtained via von Martius; sets of Wallich's later, numbered material went to both herbaria and survive today.

Some duplicates of the early Wallich collections were also sent from Calcutta in 1818 and 1819 to Sir J.E. Smith by Wallich, via Lambert, and are now in LINN, where 9 specimens of Nepalese ferns annotated as Wallich's and dated 1819, survive today and have been studied by the author. Wallich's letter of Nov. 1819 (at Kew) also requested Lambert to send duplicates of his early material to Sir Joseph Banks, President of the Royal Society (herbarium now in BM), Sir James Smith (now in LINN), Mr. E. Rudge (now in BM, see Murray, Britten & Gepp 1904), Dr. Hooker (now in K) (see Hooker *fil.* & Thomson 1855: 69), Dr. Taylor, Prof. A.P. de Candolle (now in G) and Prof. J.W. Hornemann (now in C). At about this time, Wallich himself sent some sets of his early material to William Roscoe, of Liverpool Botanic Garden (LIV), and John Lindley, of the London Horticultural Society, Turnham Green, London, and University of London. The author has worked on the fairly numerous fern-specimens from Herb. Rudge in BM, several of which have a date prior to 1820, indicating that they are early Wallich material, and thus further isolectotypes, though others are later Wallich collections with numbers. The few early Wallich specimens given to Hooker (Prof., later Sir W.J. Hooker, at Glasgow University, later Director of Kew Gardens), are now in Kew general herbarium (K), with an oval stamp on the sheets "Herbarium Hookerianum 1867" (see Hooker & Thomson 1855: 69), being part of the large Hooker herbarium (which also contained later Wallich material); they are not the ones present in K-W. Wallich also sent some such early material to the East India Company's Museum at East India House, which he later combined with the material he brought back from the Calcutta herbarium-collection to catalogue, number and distribute. All these smaller sets are duplicates of the Lambert ones and can thus be considered isolectotypes.

In addition to the main sets already mentioned, Miller (1970) showed that smaller amounts of Nepalese material from Lambert's herbarium were occasionally given to other botanists as gifts. They were mostly given out by Don, who was completely in Lambert's confidence and were presumably approved of by Lambert. Don's apparent autonomy and propriety over the Lambert herbarium, together with the fact of his being a newcomer allowed to work on Hamilton's Nepal collections, while Wallich and others were already working on Nepal, not surprisingly occasioned some adverse feelings towards him, though his position was initially unassailable out of respect for Lambert's reputation. The gifts that Lambert and he made include further isotypes of some of Don's species. Don also gave a set of Lambert's early Wallich specimens (used for the Prodromus) and probably some Hamilton ones to J.D. Prescott of Leningrad, thence to H.B. Fielding, and now in OXF (see Clokie 1964, Miller 1970, "Proctor" mentioned by Press & Shrestha 2000: 103 was in error for Prescott). Some further Nepalese ferns from Lambert's herbarium (both Hamilton's and, mostly, Wallich's) were given by Don to John Smith, Curator at Kew, whose herbarium went to the BM after he fell out with Sir Joseph Hooker (see Desmond 1966; Holttum 1967), and have been studied by the author. These generally do not have Hamilton's or Wallich's label, but are written on the sheet by Smith that he received them from Don or they were Don's original material. Some of the material of this set appears to have been somewhat muddled by Smith as to the genuine identity of some of Don's names as there are a few cases where there is clear conclusive evidence from other material that Smith was taking the Don name in the wrong sense when he wrote it on the sheets. In other cases it is difficult to be sure if it represents original material seen by Don, or was merely the same as what Smith thought a particular Don name to represent.

Other recipients of some of the material, probably including Hamilton specimens, were Prof. K.F.P. von Martius, of München (Munich), a long-term correspondent of Lambert, whose herbarium is mainly in BR, and Prof. Adrian de Jussieu, of Paris, whose herbarium is in P-JU. It appears that Don may also have given a few examples of the early Wallich specimens to Brown, as there are several ferns in BM which only have Don's labels on them and do not appear to be from any other known source. These isolectotypes, or sometimes lectotypes, are often small or scrappy specimens with little data but Napalia and the species' name in Don's hand, written in small letters and with double-slit labels that the stem had originally been inserted into.

In 1820, when his affairs were still going well, Lambert took on as librarian and herbarium curator Dr., later **Professor David Don** (1800-1841) (Don (1824), Schultes (1830), Anon. *Florist's J.* **3**: 15-19 (1842), Taylor (1842: 397-399, 478-479), Anon. *Proc. Linn. Soc.* **1**: 145-149 (1842), Luxford (1844), Stephen (1888), Don (1897), Stearn (1945, 1978), Burkill (1953), Hara, Stearn & Williams (1978), Hara (1985), Miller (1970), Stafleu & Cowan (1979), Stafleu & Mennega (2000), Smith & Fraser-Jenkins (1982), Fraser-Jenkins in prep.). There appears to be no portrait in existence of David Don; what is most probably really his brother's portrait (George Don, the younger) was published by Slack (1990), though said there to have been his father's portrait, probably in error. Don was an energetic and careful worker who played a large part in making Lambert's herbarium known to others. He published some 48 botanical papers until his life was cut short by cancer (Anon. *Proc. Linn.*

Soc. London **1**: 145-149 (1842)) of a sort commonly contracted through the smoking of clay pipes. A number of species bear the epithet *doniana* after him, such as, in ferns, *Diplazium donianum* (Mett.) Tardieu, or *Cheilanthes doniana* Fras.-Jenk & Khullar (*Aleuritopteris doniana* S.K.Wu, *nom. superfl.*; syn.: *Cheilanthes dealbata* D.Don, *non* Pursh).

David Don was born at Doo (Dove) Hillock, Forfar, on the east coast of Scotland, and died at the Linnean Society, then at Soho Square, London. He was the son of George Don (1764-1814), of a family originally from Kincardineshire, his grandfather being Alexander Don of Ireland and Forfar, Scotland (W.G. Don 1897). His father, George Don, was the Superintendent of the Royal Botanic Garden, Edinburgh, before retiring to his private plant nursery at Doo Hillock and working on his accounts of British, especially Scottish plants and those of Forfarshire, where he was well known to have discovered many rare species in the Highlands of Scotland. David Don's elder brother was George Don (1798-1856), FLS, was the first of the family to go down to London and was foreman at the Chelsea Physic Garden, London, until 1821, going on to become a distinguished botanist who described many species of flowering plants from Britain and the tropics, being a collector for the Horticultural Society of London in Central and South America and West Africa. His younger brothers, Patrick and James, were also gardeners, at the famous private gardens at Bedgebury, Kent, and Knowle Park, Kent, respectively.

As a young man, David had trained as a horticulturalist under his father, then continued training in employment at Dickson's Nursery, Broughton, Edinburgh. His first appointment was in 1819, when he followed his elder brother and moved down to London, to join the horticultural staff at the Chelsea Physic Garden (then known as the Apothecaries' Garden), one of the oldest Horticultural gardens in Europe, still thriving today under the aegis of the Natural History Museum, South Kensington. But with his father's introduction to and the subsequent recommendation of Robert Brown of the Linnean Society (the distinguished collector with Sir Joseph Banks in Australia, and author of *Prodromus Floræ Novæ Hollandiæ* (1810) and other works), he was soon taken up by Lambert in 1820 and lived in Lambert's London house, converted into the botanical museum, herbarium and library, remaining his librarian until 1836. A glowing first-hand account of what it was like to visit Lambert's herbarium and meet the most dedicated and obliging David Don was written by a visiting German Professor in 1824 (Schultes 1830). While maintaining his work for Lambert, he also became Librarian to the Linnean Society of London, nearby at Soho Square, London, in 1822, succeeding Brown on his retirement, and after 1836 residing there with his wife. He was then apparently no longer able to stay in Lambert's household due to having married one of his girl-servants without approval (Gray 1893, Miller 1970). He remained in the Linnean Society post while going on to become Professor of Botany at King's College, London, in 1836. At the time of his early death from cancer of the lip and then face and neck, he was an established and much respected and liked botanical authority, whose considerable body of work, as well as a reputation for helpfulness, modesty and good sense had become internationally known. He was buried at St. Agnes Cemetery, Kensal Green, London, after a funeral attended by most of the distinguished English botanists of the day (Anon. *Florist's J.* **3**: 15-19 (1842)). He and his wife had no issue. Bibliographical details of his nearly 50 publications are given in Anon. *Proc. Linn. Soc. Lond.* **1**: 145-149 (1842) and by Stafleu & Cowan (1979) and Stafleu & Mennega (2000), where it can be seen that apart from monographs of several genera with some Nepalese species (*e.g.* Don 1825, 1841), he also produced one other paper mainly relevant to Nepal (Don 1820), though none of them give significantly more precise details as to localities or collections.

Don was commissioned by Lambert, with the approval of Hamilton, to make a preliminary account of the otherwise unstudied Nepalese collections of Hamilton in the Lambert herbarium. To these he also added the early Nepalese collections sent to Lambert by Wallich, though this was not approved of by Wallich himself, but it must have been known to Lambert and others could hardly complain against Lambert's plans. Hamilton also informed Wallich of Don's work on both their collections. However Don was not given access for his *Prodromus* (though he saw it later, Don 1841) to Smith's material from either Wallich or Hamilton, including Hamilton's manuscript of *Flora Nepalensis*, which was all kept personally by Smith (and not even made available for Hamilton), despite Don's being Librarian to the Linnean Society (Don 1824, preface, Prain 1905). A letter of Hamilton's to Wallich of 16 Oct. 1821, cited by Prain and Stearn, said "A Mr. Don, however, who lives with Mr. Lambert, to whom I gave duplicates of the collection presented to Sir J.E. Smith, is engaged in publishing an account of them together with those which you have sent, and I believe he has both abilities and industry to produce a very valuable work. Whether or not Sir J.E. Smith will allow him the use of my drawings and written descriptions I have not learned." It is clear that Smith did not allow others access to his Nepalese material, though he was unable to publish on it himself to any worthwhile degree. It is conceivable that this might have some bearing on the unexplained loss of much of the material Smith was given.

Don began this work in 1820 and took 5 years to complete it (Stearn 1945), publishing it in December 1824 as *Prodromus Floræ Nepalensis, sive Enumeratio Vegetabilium quæ in itinere per Nepaliam proprie dictam et regiones conterminas, ann. 1802-1803, detexit atque legit D. D. Franciscus Hamilton* (*olim Buchanan*) *M.D. Societ. Reg. et Linnæan. Londin. Soc. Accedunt plantæ a D. Wallich nuperius missæ, secundum methodi naturalis normam disposuit atque descripsit*, and dedicating it to the Societas Mercatorum Indiae Orientalis, or East India Company. Stearn (1945) has concluded that because Lindley stated it was published in February 1825, and the work is dated on the title page as 1825, "it was obviously not intended or expected to be available to the public until 1825", so there seemed no reason not to accept Lindley's date. However Don himself (1841: 518) stated that "it was completed and some copies of the work distributed before the close of 1824" and as the distribution of copies effects publication under the *International Code of Botanical Nomenclature* (2000), the date of Dec. 1824 is accepted here. Although only a Prodromus, not purporting to be a complete Flora, Don treated nearly 700 species, fifty of them from India (mostly from the adjacent region of Kumaon, previously held by the Nepalese), including 87 pteridophytes (32 collected by Hamilton, 63 otherwise or also collected by Wallich (including 1 unattributed specimen, *Peranema*) and 1 by Kamroop, from

Kumaon). It was thus an impressive, if highly incomplete start to documenting the Nepalese flora and though not his first work (he had published a few other papers on Nepal *etc.*, starting in 1820) was made while he was still somewhat inexperienced, making a remarkable early landmark in his botanical career.

The great importance of *Prodromus Floræ Nepalensis* today is that it was published at an early date well before the mid-century works of the great investigation of Indian subcontinental botany at Kew, Berlin *etc.*, and also before Blume's (1828) work on Java and Wallich's on the Indian subcontinent. It predates Presl, Hooker and his colleagues, Kunze, Roxburgh's ferns in Griffith (but not, of course, Roxburgh's main works) and many others. Don's names thus have priority over many others' in Asia, being one of the earlier works in that continent, and it is therefore of great importance to find out exactly what he meant by a name. Don's nomenclature and the typification of his species have important repercussions throughout the whole of Asia. Yet because of the dispersal and near loss of Lambert's herbarium and the apparent loss of relevant portions of Smith's, along with the simple descriptions, the identity of Don's species has often been badly confused and misunderstood. Too high a proportion of his names have been ignored as unidentified. In some respects, if Lambert's was a "lost herbarium", Don's is a partly "lost Flora". To add to the difficulty, though he usually gave actual localities for Hamilton's specimens, he could not do so for Gardner's as Wallich did not provide any more detailed locality for them. Don usually cited the material for the bulk of these collections as merely, "*Hab*. in Nepaliâ. *Wallich*", or "*Hab*. in Nepaliæ alpibus. *Wallich*". This has meant that the only guide to identifying his species when based on Wallichian collections is his usually brief and often general description, combined with the external evidence as to how authorities such as Sir W.J. Hooker (1785-1865), Thomas Moore (1821-1887), and later R.H. Beddome (1830-1911) and C.B. Clarke (1832-1906), took the name.

Some of the most obscure of Don's fern-names, or those with missing types, have become clear mainly due to one important exception to Don's work being "sent into Coventry", namely **Thomas Moore's** (1857-1862 and *ined.*) *Index Filicum*. Thomas Moore was the distinguished pteridologist and Curator of the Chelsea Physic Garden, London, from 1848-1887 (Stafleu & Cowan (1981), Laird (1988), Desmond (1994)). His great work gave more detail than any later Indexes and Moore took the trouble to investigate Don's names, but unfortunately the publication of the Index was not completed for financial reasons and it only goes down as far as the letter G (*Goniophlebium*). But letters G-Z were also written in manuscript by him and a paper by Underwood (1905) suggested the possible existence of the missing section at Kew (Fraser-Jenkins 2004). At the request of the present author, the bound manuscript of this "lost book" has been rediscovered in a perfect state of preservation by the archivist at the Royal Botanic Gardens, Kew. On study by the author, it was extremely pleasing to find that Moore gave reference to all of Don's names, most, but not all of which he identified as a result of his near contemporary knowledge, including his knowing Don during his lifetime. The unexpected rediscovery of this work after a gap of 150 years is of considerable interest and the present author has requested to have it microfiched with a view to working further on it and making it more widely available. Several of Don's most difficult fern names have been identifiable again as a result of information and synonymy given by Moore.

Don's work profusely acknowledges the collections of both Hamilton and Wallich, together with the rôle of the East India Company, to whom it is dedicated, and who would probably have been pleased in principle (if not as to detail, according to Lindley 1825) to see some preliminary overall account relating to their Indian (subcontinent) collections published quickly. It can hardly be denied that the slower pace of plant-description in Wallich's vast and more detailed work and its concentration on detailing only a few species in sumptuous and expensively exclusive format (apart from his later compilation of a bare list of undescribed names in his unpublished Catalogue, or *Numerical List etc.*) had left the bulk of the collecting activity by Wallich and all his colleagues in the area unsung and unaccounted for for a number of years. However, as soon as Don's work was published it became clear that it was a rather superficial work. The descriptions were too brief, often did not hit upon the essential diagnostic characteristics, and were poorly or inadequately written, by the standards of the time as well as today. A number of species were also generically misplaced by the concepts of the day, as now. As Don intended, it was only a Prodromus, but unfortunately it then became the subject, along with its author, of some of the severest hostile criticism ever published in botany by some of those whose work it was unwittingly pre-empting, Lindley, Wallich and a more temperate comment by Hamilton himself. It became something of a blackballed work, mostly avoided by the early Victorian botanists unless it really had to be quoted (a situation the present author is not entirely unfamiliar with himself; see Khullar 2000: 542-544, in reaction to his analysis (Fraser-Jenkins 1997) of the serious problems inherent in modern Indian pteridology, though at least the former critic has now come to agree on most taxonomic points!). All this only contributed to making Don's work more obscure, since contemporary authors who may well have known the identity of many of the names tended not to mention them. Hooker and others often preferred to mention only those of Wallich, even when those were merely undescribed *nomina nuda* attached to the specimens, as at least those from Wallich's *Numerical List* were thought to have priority under the practice prevalent at the time. In a few cases where types appear to be missing in the BM, the avoidance of Don's *Prodromus* has contributed towards making the status of such names equivalent to the old concept of *nomina dubia*, though no such nomenclatural catogory is permissible today under the modern Codes of nomenclature. However the successful modern procedures of lectotypification and neotypification (also of epitypification if necessary) serve to remove all ambiguity as well as to make secure decisions which it is compulsory to accept and which cannot be changed without specific reasons complying with the Articles of the Code. The present author is currently engaged in applying these procedures in a detailed typification of Don's pteridophytes (Fraser-Jenkins, in prep.).

Prof. John Lindley FRS, the orchidologist, of the London Horticultural Society and University of London, took up cudgels against Don (Lindley 1825), as it were on behalf of Wallich, and hinted that Lambert himself should not have allowed it to happen. Details of his extraordinary criticism, which though partly justified, can be seen today as being considerably overstated and not making allowance for the situation at the time of writing the book, were given by Stearn (1945).

Lindley commented that "the general character of that work is not such as to entitle the author of it to our credence. Under the pretence of publishing the collection of Nepalese plants, formed by Dr. Hamilton during his residence in India, the writer of this book has had the extraordinary assurance to incorporate with them a considerable proportion of the plants collected in Nepal by Dr. Wallich, by him confided to individuals in this country, and at this moment publishing.... What object can have been expected by this production we are unable to understand ..... We do not suppose the Honourable Court of Directors of the East India Company, to whom this production is dedicated, will feel much flattered." He also complained that it was "written in so strange a Language, that we can scarcely guess at its name, unless, indeed, it be a specimen of some new kind of Latin ..." written "with great facility, after three lessons of an hour each, without the incumbrance of previous education". However part of Lindley's fiery criticism was undoubtledly based on truth, as in reality Don was merely giving brief descriptions of the most common species, and not even all of the 107 Nepalese pteridophytes of Lambert's herbarium that he quoted previously (Don 1828), as can also be seen from the original material at the BM, which contains several obvious species of both Hamilton's and Wallich's that he overlooked.

More seriously Lindley picked out the work's major drawback, of considerable import for today's workers, that Don did not refer to Wallich's manuscript names, though this was undoubtedly because the early collections had been sent to Lambert without names and this was the situation at the time Don started work. The surviving specimens themselves only bear later Wallichian names written on them in pencil probably after their acquisition by the BM. But many of the herbarium-names Wallich gave thereafter to his collections became known to a number of other botanists Wallich had sent material to before his Catalogue was made available, and the existence of these names was sometimes in time for Don to have incorporated them in the Prodromus, had he thought to add them in to his account and could he have anticipated how important they would become as Wallich's work proceeded. This has become yet more serious today due to the fact that his species cannot be related to Wallich's later numbered set, which was not enumerated until Wallich's return to Britain in 1828, as the Prodromus was based on the earlier collections and the Wallich (1828-1849) numbers did not exist at the time. All this has rendered it extremely difficult in a number of cases to identify which species Don could have been referring to. Lindley remonstrated "why the numerous species from Dr. Hamilton himself, already extremely well published [but in fact, not so, and anyway excluding Pteridophytes] by the learned President of the Linnean Society, should ..... never have been referred to [though as we now know, this was because Don was not given access to them]; upon what pretence the published names of Wallich, de Candolle, and others, should have been, in numerous instances, altered; and why, of the Nepalese plants in the Herbaria in London, few besides species of the most obvious character and easy determination should have been selected for the display of Mr. Don's learning and abilities; these are all points well deserving the consideration of Botanists". In fact we know that relatively few species were described because this was nearly all the material available to Don at the time from Hamilton's and Wallich's early sets. Lindley also included a thinly veiled criticism of the respected Lambert himself, who, after all, had initiated the work, complaining "that men of sense and character should be found in this country ready to countenance such a feeling [that Don vainly supposed himself more competent than the Indian botanists, *i.e.* Wallich and Hamilton, themselves], is most incredible and ridiculous". In fact the exaggerated tone of Lindley's attack, seen today, tends to reflect more adversely on its writer than on the subject of the criticism!

Yet from one or two inexplicably bad instances of near plagiarism in Don's book, it is not surprising that the normally most personable and kindly Wallich himself hit the botanical roof with abundant insulted vigour! He responded furiously to Don's book in his *Plantæ Asiaticæ Rariores* (1829-1832), under Plate 48 (1830: 41-42, *Sphaeropteris*), "It is necessary to explain the reasons for my having preferred a manuscript name for my fern [*Sphaeropteris barbata* Wall., now called *Peranema cyatheoides* D.Don] to one already published. It is eleven years since I first furnished my correspondents in England with specimens of it; and it was included in a general collection of ferns, which I sent in 1823 to the Museum at India House. Of the arrival of this collection I presume the author of *Prodromus Floræ Nepalensis* cannot have been ignorant, since the Hon. East India Company most liberally distributed the duplicates in May of the following year. In the Museum just mentioned the fern was labelled, "*Cyathea*? *barbata*, Wall., *vix non genus novum*, *Sphaeropteris Nob*., *non* Bernhardi" [Wallich's generic name thus clearly being an unuseable later homonym] a name which Mr. Brown did me the honour of adopting immediately, in preference to that of *Poedilema*, which he had given the plant, at the time he procured the beautiful drawing, from which the details of my plate were derived. In the work alluded to, neither is any notice taken of the name proposed by me, nor is even the source mentioned from whence the fern was obtained; and in a number of other cases the author has treated me with still less courtesy, inasmuch as he repeatedly quotes my names of plants, and at the same time rejects them. In the instance of *Aconitum ferox* (see above p. 35) he cites that name, he even repeats it, as having been adopted by my illustrious friend Professor DeCandolle, from my manuscript [Wallich's *Filicologia Nepalensis* (1821, *ined*., see below and see Fraser-Jenkins in prep.]; and yet he rejects us both, and introduces the species under the name of *A. virosum*. In the same manner he disposes of an *Aspidium* [*A. wallichii* Hook., *non A. wallichianum* Spreng., which Don named *Neuronia asplenioides* D.Don, *nom. superfl*., presumably because he erroneously thought, as did others later, that *wallichii* and *wallichianum* were homonyms, now known as *Oleandra wallichii* (Hook.) C.Presl] from Nipal, called after me, and published by a botanist of the greatest authority and celebrity, one of the dearest and most esteemed friends I possess in the world, Professor Hooker; not to mention a great many other instances of a similar nature, and some of them still more pointed, which occur in his own publication, as well as in his additions to those of others. Having never in my life experienced such conduct, or any thing in the remotest degree like it, from any other quarter, I might have been justified in the course I have taken; but there is another reason for my not adopting the name *Peranema*, which is, that there exists already a genus of *Verbenaceæ* called *Peronema* by the late Mr. Jack, in a work which was published in 1822 (*Malayan Miscellanies, vol*. 2. *no*.7 *p*. 46). That these names are likely to be confounded must at once be evident, and is proved by M. Kaulfuss, who in speaking of this genus of ferns uniformly calls it *Peronema*." In fact the two names are not orthographic variants and thus not homonyms, but there is no reason beyond a certain independence, why Don should have ignored Wallich's name.

Finally Prain (1905: 32) revealed that Hamilton's reaction, though somewhat milder, was also one of partial dissatisfaction, in his letter of 10 Feb. 1826 to Wallich, concerning the ongoing work on his material he said "Brown, I understand has become very lazy and Smith is nearly *ab agendo*. Don, I believe, is at present the most active botanist about London, but has not yet acquired sufficient experience to render him at all correct. His Prodromus Floræ Nepalensis abounds in errors".

Poor Don! With hindsight, today, it can be seen that apart from the error of not using those of Wallich's names that were already known, much of the problem actually lay not with Don, but with the circumstances of the collections available to him, as much as with Lambert's independent initiative. It was Wallich's own poor localisation and, at that time, lack of numbering, that are reflected in the *Prodromus*, plus the fact that the collection was not made by Wallich himself but by a less botanically minded amateur, Gardner, and being at an early stage of his programme, was both basic and incomplete. It can be seen that the *Prodromus* was indeed just that, a preliminary account, produced for Lambert at too early a time compared to what was to come, after 1828, from the delayed release of the more substantial material from Wallich's own visit to Nepal in 1820-1821. However, notwithstanding the furore over this early work, Don's career later went on to aspire to considerable heights. Indeed, quite soon after this work he was able to rehabilitate his reputation and gain considerable standing in the botanical society of the day. By the time of his death he was widely and highly regarded for his considerable body of thorough published work, and his character was described by Richard Taylor (1842), co-editor with him of *The Annals and Magazine of Natural History*, as "unpretending, disinterested [i.e. unselfish, not following self-interest], openhearted, and sincere. His native kindness, cordiality and hilarity as a companion will long be affectionately remembered by those who knew him".

The material Don used which has been called here the "early Wallich material", dated from 1817-1819 (and perhaps some also collected in

1820, but prior to Wallich's visit to Nepal) was not in fact collected by Wallich, who did not go to Nepal until 1820. It bears his name because he included it in his herbarium-sendings and its collector was not a professional botanist, so was not named on the specimens - hardly a practice that would be followed today. The material was actually collected by **The Hon. Edward Gardner** (1784-1861), the first permanent British Resident (or Ambassador) in Nepal, and the second botanical collector in Nepal, it is therefore recommended to cite the early Wallich material as being collected by "Gardner in Wallich, 17 March 1818" *etc.* Gardner's important contribution to Nepalese botany has been much overlooked and his name and details of his life are virtually unknown in this field. For example, Hara (1985) did not know of him and appears to have thought Wallich himself had collected in Nepal in 1818. Despite being well known historically he has been most unfortunately mistreated by the tides of botanical history and has become the "lost botanist" of Nepal. Gardner was the fifth son of the distinguished Admiral of the Blue Sir Alan Gardner (1742-1809), the 1st. Baron Gardner, of Uttoxeter, England, and Commander of his Majesty's ships in Jamaica (Hooker (1820): *sub t.* 146; Debrett (1825); Wallich (1830); Hunter (1896: 58-59); Landon (1928); Burke (1967: 1012-1014); Phillimore (1950, 1954); Whelpton (1991); Khanduri (1997); Gould (1999); Dalrymple (2002); letters to and from Wallich from 1817-1846, were given to CAL, but only those from 1831 onwards and (inexplicably) the letters of the Rev. William Allport Leighton (1805-1889), found there by the present author; letters to Edward Gardner from W.L. Gardner from 1820-1821 in Indian Office Library, London, with transcripts in National Army Museum, Chelsea, London (kindly provided to the present author by W. Dalrymple)). No portrait of Edward Gardner is known to exist, though his father's and cousin's portraits are known; but he was graphically, indeed dramatically described by Lady Maria Nugent (1839), quoted by Dalrymple (2002: 53-54), when she was indignantly complaining about her fellow countrymen who had unashamedly "gone native", "I shall now say a few words of Messrs. Gardner and Fraser who are still of our party. They both wear immense whiskers, and neither will eat beef or pork, being as much Hindoos as Christians, if not more; they are both of them clever and intelligent, but eccentric; and having come to this country early, they have formed opinions and prejudices that make them almost natives. I endeavour to insinuate every thing that I think will have any weight with them. I talk of the religion they were brought up in, and of their friends, who would be astonished and shocked at their whiskers, beards, *&c. &c.* All this we generally debated between us and I still hope they will think of it." Fortunately it seems clear that they were able to resist this veritable battle-axe of a neo-Christian British Empire-builder!

Edward Gardner went to India as a secretary, or Writer, in 1802, when he was 18, becoming a registrar and Assistant Magistrate at Aligarh, now in Uttar Pradesh; but in 1808 he was advanced to Assistant Resident at Delhi, and five years later became Acting Judge and Magistrate at Moradabad, in northern U.P. His first experience with the Nepalese was when he was posted as Commissioner and Political Agent to the Governor-General of Bengal, Francis, Earl of Moira, the future 1st Marquess of Hastings, in Kumaon at the outbreak of the Gorkha War, in 1814. Hunter (1896) described him as the Marquess' right-hand man in bringing Nepal into treaty-relations with the British in India. He has often been confused with his cousin, W.L. Gardner, for example being described by Hooker (and Smith & Parsons 1970) as an army Colonel, which was not the case.

He was a cousin of and frequent correspondent with the remarkable Lt.-Col. William Linnaeus Gardner (1770-1835) (see Stephen (1890); Hunter (1896); Burke (1967), Stiller (1973); Gould (1999), Dalrymple (2002: 141 *etc.*, 2003); portrait in Dalrymple (2002)), of Khasganj, north-east of Agra, in Etah District, Uttar Pradesh, where his family descendant, Julian Gardner, proven legitimate heir to the Barony of Gardner, and his family still live today, speaking Hindi and living an Indian village life as peasant-farmers among their family memorials (Dalrymple 2003). William Gardner was the fearless hero of the British battles in Kumaon, who, along with Edward Gardner, was selected by the Marquess of Hastings to spearhead the British campaign against the Gorkha Rajahs' expansion into Kumaon, Garhwal, Sirmoor and beyond Simla. William Gardner was the son of Major Valentine Gardner, of the 16th Foot Regiment, in America during the American War of Independence, elder brother of the 1st. Lord Gardner, and was thus Edward Gardner's first cousin. He was born in Livingstone Manor, New York State, and was a godson of Carl Linné (Linnaeus), the Swedish Botanist-founder of modern Biological nomenclature - i.e. Linnaeus was a family-friend chosen to oversee his religious upbringing. After the final British collapse in America in 1782, he fled from America to India, arriving the next year to join the regular Indian Army in 1789, being appointed a Captain, then in 1794 retiring to join the army of the Nizam of Hyderabad, followed by the Army of the Marathas. In 1803 he rejoined the British and founded an irregular cavalry known as Gardner's Horse or Gardner's Irregulars (the 2nd. local horse, later becoming the 2nd. Bengal Cavalry, after their loyalty to the British in the Indian Mutiny of 1857), later becoming promoted to Lieutenant-Colonel. His independent Company which was assigned to Almora in the Kumaon Hills under Col., later General Sir David Ochterlony, at the same time as Edward Gardner was making political moves in the area, and succeeded spectacularly against the Gorkha army, turning the tide of the war, when all the other British regular officers and Generals had fallen apart in disarray after loosing heavily to often considerably smaller and underarmed Gorkha detachments. But after he had taken Almora in 1815, due to his outstanding strategy, he was nevertheless relieved of charge by Col., later General Jasper Nicolls and it was widely felt that his irregular mercenary and Indianised background had been held against him unfairly. He had become Muslim and married "Her Highness Furzund Azeza Azubdeh-tul Arrakeen Umdehtul Asateen Nuwab Mah Munzil ul Nissa Begum Dehlmi" (Parkes 1850, cited by Dalrymple 2002), otherwise known as Begum Munzel ul-Nissa, the fourteen year old daughter of the Nawab of Cambay (Dalrymple 2002), and lived in Indian style with his multiplying family.

Edward Gardner also had the idea to pioneer the first recruitment of defecting Gorkhalis and Kumaonis, whom the British had been much impressed by, into the British Army as Gorkha Battalions, which he did together with his cousin, and made it a formal process after 1825, which was further enhanced by his successor in Kathmandu, Brian Houghton Hodgson, the tradition continuing to the present day. Due to his considerable abilities the Marquess of Hastings, the Governor-General of India from 1813, then selected Edward on merit over the heads of his seniors to represent British interests as Resident, or Honorary Consul, to the Court of the Rajah in Kathmandu. His position had been created as one of the conditions of the Treaty of Sugauli (a village near the modern town of Roxaul, Bihar) with the Gorkhas in March 1816, that brought the Anglo-Nepalese "Gorkha war" of 1814-1816 to an end, defined the borders of an independent, modern Nepal, which the British themselves helped to maintain and did not subsequently interfere with, and kept British visitors and interests out of Nepal, apart from the agreed presence of the Resident or Ambassador there. Before Gardner arrived in Kathmandu, his position was held by an acting Resident, Lieutenant John Peter Boileau, also in action in Kumaon in 1814-15, after whose cousin, the surveyor, Lt. A.H.E. Boileau, Boileauganj, below Mussoorie, was named (though now often corrupted to "Boilerganj" due to the laundry, or *dhobi-wallahs* there!). Boileau reached Kathmandu as the commandant of the Resident's Escort and Postmaster to the Residency, a far from easy task as the Residency was deliberately kept isolated by the Nepalese, in April 1816 and was relieved by Gardner later that year, but died of malaria at Patna, while returning towards Calcutta via Muzaffarpur.

Gardner, who remained Resident for 14 years until 1829 (apart from some months away as Resident in Bundelkund and Superintendent in Narbada, Central India), had a brief to attempt to convert the climate of hostility into one of friendship, or at least stability in mutual respect. But given the continuing strong post-war hostility of the defeated Nepalese to the British throughout the first part of his time there, this was at first an almost hopeless task, though he largely succeeded later. Like Hamilton and Knox, Gardner was again stationed in somewhat constrained conditions, though they improved later after Lord Hastings decided to return the Nepal *terai* to Nepal under its ruling Prime Minister, Gen. Bhim Sen Thapa. In fact Nepalese *sepoys* (soldiers) were stationed between the Residence and Kathmandu to intercept all messages from the town and arrest their bearers, and all contact between Nepalese and the Residence was strictly prohibited, as was any contact between the Residence and British traders to Kathmandu (Hasrat 1970). It is fortunate that Gardner's letters to Wallich were allowed through as diplomatic correspondence. It was during long unproductive periods of waiting that he turned to his new interest in Botany. With the help of his Assistant Resident, Robert Stuart, he directed a small team of assistants to make collections, many of which he made himself, in 1817, 1818 and 1819 (and possibly also in early 1820 when Wallich himself came to Nepal). He also created a garden in the Residency grounds, where he planted a number of meritable Nepalese trees and shrubs and from which some of the material sent to Wallich was obtained, labelled as *in hort*. The garden-park surviving there today appears to be the one he founded, though the grounds are now divided between the British and Indian Embassies (since Indian Independence in 1947, when land was donated by the British Embassy).

He was succeeded as Acting Resident from 1829-1831 by the famous **Sir Brian Houghton Hodgson** (1799-1894), who initially trained under him as his Assistant Resident, from 1820-1822 and 1824-1829. Hodgson wrote glowingly of Gardner, "Another man to form myself after - a man with all the simplicity and more than courtesy of the Commissioner of Kumaon [George Traill, whom he had been stationed under] and a man who was the perfection of good sense and good temper - a man who liking the Nepalese and understanding them was doing wonders in reconciling a Court of Chinese proclivities to that offensive novelty of responsible international dealing through a permanent diplomatic establishment in their midst" (Waterhouse 2004: 4-5). Hodgson himself played an exceptional and constructive diplomatic rôle in Nepal's history until he retired in 1843, living in Darjeeling from 1845-1858 writing scientific papers and finally returning to Britain. Apart from his celebrated rediscovery of the texts and structure of the Buddhist religion, Hodgson also became the founder of Himalayan zoology, publishing a great number of outstanding scientific papers (see Hunter 1896, Landon 1928 **1**: 272-284; Waterhouse 2004) and sending all his cultural and scientific material and manuscripts for publication down to Wallich in Calcutta for onward transmission. He also supervised a couple of collectors for Wallich for a number of years after his visit, collected occasional botanical specimens for him and wrote abundant letters to him about his discoveries in many fields (CAL!). One can but imagine Gardner and Hodgson working on their new findings on all possible occasions when they could get away from duty and the excitement of unearthing many of the botanical and zoological treasures of Nepal (and indeed the whole Himalayan region) for the first time! After a brief interlude, when Sir H. Maddock was appointed as Resident from 1831-1833, Hodgson became Resident from 1833-1843 and became a confidant of Nepal's famous early Statesman-Prime Minister, Bhim Sen Thapa, thus founding successful relations between the British and Nepalese, which he was able to maintain during both calm and crisis throughout his long service.

As a passionate botanical collector, Gardner sent Wallich numerous packages of herbarium-specimens as well as a good number of living plants and these specimens (along with those of de Silva and Singh, working under the Residency team) are what are here referred to as the early Wallich herbarium collections. In 1817, the year of his official appointment at the Garden, Wallich had deputed two of the collectors working under him to visit Nepal. One was **Francis de Silva**, who also collected for Wallich around Sylhet (now in Bangladesh) and in the Khasi Hills (now in Meghalaya State, India); the other was **Bharat Singh**, another collector from the Garden (Desmond 1992: 132). The Index to Wallich's incoming and outgoing correspondence held in one of two archives cupboards in the library of the Central National Herbarium at Sibhpur, Calcutta, lists a letter of 22 Feb., 1819, from de Silva from Kathmandu with a list of products from Nepal, where he was described by Wallich as "Plant collector in Nepal". As mentioned by Wallich in a letter of 13 Oct. 1818 to Sir W.J. Hooker (*Kew Director's correspondence* **52**: *f.* 47), Gardner was not a trained botanist himself, but sent in "invaluable treasures". He was nevertheless a very enthusiastic and knowledgeable amateur botanist in his spare time - of which there seems to have been plenty. In addition, though Waterhouse (2004: 4) thought Gardner was not allowed to travel outside Kathmandu Valley, he was evidently slightly less restricted than either Hamilton or Wallich as he was able to make at least one visit outside the valley, up to the high ranges of Gossainthan north of the valley, in 1818. This is shown in the letters of Wallich to W.J. Hooker, of 2 Sept. 1818 and 13 Oct. 1818 in the *Director's Correspondence* vol. **52**, in the archives at Kew, where it is mentioned that Gardner had just been sent on an expedition to the Snowy Mountains, or Himalaya, and that he had by then sent Wallich upward of 1,000 species (the author is most indebted to Prof. D. Arnold for kindly providing this information). A letter dated 14 Oct. 1818 - from "The Honb. E. Gardner, Nipal, with list of Stages from Katmandoo to

Gossain Than, and remarks on the route" is listed in the Index to Wallich's correspondence at CAL (!) and must presumably refer to the same visit, though the present author has not yet been able to find the letter itself in CAL; Gardner's locality Gossainthan is also mentioned by Wallich (1830: 35). In 1817 Gardner sent Wallich specimens from Nuwakot ("Noakote"), outside the Valley to the N.W., but it appears that the specimens were brought to him by one of the collectors (Wallich 1830: 32-33). He wrote informed details of his most interesting findings in his very many letters to Wallich. At one stage he was writing up to some 5 or 6 letters each week, and wrote at least every week for a period of about six years from 1817-1823. After Feb. 1824 his letters decreased considerably, at which time his Assistant Resident, Hodgson, began his correspondence with Wallich, according to the Index of Wallich's correspondence. Gardner also wrote occasional letters up until 1861 to Wallich after his retirement in 1829 to live at 20, Bruton Street, London.

It was deservedly in his honour that a number of his collections of new species were named after him by Wallich and others (including *Coelogyne gardneriana* Wall., *Daphne gardneri* Wall., *Hedychium gardnerianum* Wall., and the moss, *Calymperes gardneri* Hook. (1820)); Wallich noted under *Gardneria ovata* Wall., List no. 403 "I have dedicated this new genus to the Hon. Edward Gardner, late Resident at the Court of the Rajah of Nepala, by whose indefatigable exertions & liberality the Botanic Garden at Calcutta has been enriched by the most splendid additions of flowering plants, trees, and specimens of the interesting & hitherto mostly unknown vegetable productions of that country". As can be seen from his collections in the BM, he sent some material to Wallich in 1817, with a larger lot in 1818 and 1819. Along with the specimens he also gave Wallich information about their vernacular names (written on their labels by Wallich), but surprisingly this is not in Nepali, as none of the names correspond at all with the Nepali names for the various ferns. It is also not in Hindi or Bengali and appears either to have been Newari, the main tribe in the Kathmandu Valley (in which case such names have now generally been forgotten by modern day Newars in Kathmandu and elsewhere), or perhaps Sherpali or one of the other Tibeto-Burman tribal languages, such as Tamang, Gurung or Magar. The author has not yet been able from his enquiries in Nepal to ascertain which language they are in or to find anyone who can recognise the general gist of the names.

Working and collecting with Gardner was his first Assistant Resident, **Robert Stuart** (d. 1820), who also sent Wallich many botanical and other letters including some lists of plants. Stuart's tall funeral monument in the British Embassy graveyard (founded 1816), at Samakushi, just below the beautiful forested ridge on which the Embassy still stands today as in Gardner's time, is in a good state of preserve today, and reads (in full) "To The Memory of Robert Stuart , Esqre. Third son of Sir John Stuart, Bt. of Allanbank in North Britain And Assistant to the first British Resident at the Court of the Raja of Nipaul, who died at this Capital on the 14th of March, 1820. A few of his friends Have recorded on this Monument A slight and sincere Though melancholy Testimony of the Ability, Excellence, and worth, Which, while he lived Commanded their Esteem, Love, and Respect And now that he has been taken from them Connect the remembrance of him With their warmest feelings. Simpson & Llewellyn Setr.". He caught fever (presumably malaria) while visiting the Nepal *terai* in order to accompany Wallich to Kathmandu, and one of the mentioned friends was Wallich himself, who helped arrange for the monument and recorded his sympathy in a note on the last letter he received from him, dated 8 Feb. 1820 (in the Index to Wallich's letters, at CAL!). The present author has not yet been able to find out where and when Gardner died, though his letters show he was still in Nepal in 1825, when he would have been 51 and he retired as resident four years later. Gardner's earlier letters to Wallich (from 1818 to 1821) have not yet turned up during the present author's searches at Calcutta Botanic Garden, though they may still survive in some of the locked cupboards seen by him in the disused old library there. It is hoped to search further for them as the Index shows that they contain details of his collections and sendings. A further source for details of Gardner's excursions and activities is the official log of the Nepal Residency, preserved in the Residency papers at ref. R/5 in the India Office section of the British Library, London. It is hoped that some details of Gardner's excursions and collection-localities may be found there if the relevant Gardner letters fail to turn up at Calcutta.

Despite Gardner's success in Nepalese collection, it is a surprising accident of botanical history that few modern workers on the region properly accredit the collections to him or even seem to be aware that it was entirely due to him and his team that Wallich's early Nepalese collections were made. Instead most workers refer only to Wallich as if he were the collector and it seemed that Wallich himself almost monopolised some of the more minor collections of the many botanists that he brought back to Britain from Calcutta to catalogue and work through, even though he normally cited the more well known collectors' names. The present author recalls once being mistakenly asked in India in the 1970s how it was that Wallich was able to make Nepalese collections in 1819, and whether he might have made a clandestine cross-border visit to Nepal prior to his documented visit a couple of years later - it must be said that he also recalls his being unable to explain it at that time, except to think that he might have received specimens from others! It is hoped that this present paper will lead to a more clear understanding of Gardner's great contributions to Nepalese and Himalayan botany, as the second botanist there after Hamilton.

The next botanical visitor to Nepal was **Lt. William Jack**, MA., MD, Surgeon (1795-1822), eldest son of the Revd. Prof. William Jack, MA, MD, Professor of Mathematics and Philosophy, and Principal of the University of Aberdeen. While serving with the Military in the Gorkha War, he was able to make a number of collections, but no ferns (Jack 1835, Gage & Burkill 1916); his surviving herbarium being in K-W, K, BM, E and G. Jack was sent to join the column advancing under General Sir David Ochterlony towards Kathmandu via central Nepal during the Gorkha War, in 1815, and visited the Churia Ghat Hills in January that year, then to the camp at Bechiako (Bichakori, in the dry river bed south of Nayabasti, a few km. south of the Churia Ghat pass, leading over to the Rapti Valley and Hetauda), returning to Dinapur, Patna in May 1815. He collected and described flowering plants from lower Nepal, but apparently no ferns, his unpublished material being given to Wallich, who published a few of his species (Wallich 1818), and with whom he stayed at Garden Reach, near the Botanical Garden, Howrah. He had already become ill with probable Tuberculosis, contracted during his visit to Nepal, but in Dec. 1818 went to Penang and Sumatra with Sir Stamford Raffles, the founder of Singapore, where he made good collections and discovered many novelties, which he wrote about in his letters, corresponding with many well known botanists of the day, including Lambert. Finally he caught severe malaria in Sumatra and was sent from his home in Bencoolen, Sumatra, on a voyage to Java, hoping to recover, but his weakened constitution could not recover and he died the morning before departing for Cape Town. Unfortunately his Malesian herbarium and papers, along with the extensive unpublished papers of Sir Stamford Raffles were completely destroyed when the ship 'Fame', on which the Raffles's were returning to England, caught fire, exploded and sank in 1824, as described by Sir Stamford Raffles (Gage & Burkill 1916: 239-241). Jack is not to be confused with a later William Jack, staying at Garden Reach, Calcutta, in 1834, whose letters to Wallich concerning his sinking health due to malaria are at CAL (!).

Our final early botanist, actually the fourth to visit and collect in Nepal, is the botanical colossus of a figure, **Dr. Nathaniel Wallich**, MD, Ph D, FRS, FLS, Knight of Dannebrog *etc.* (1786-1854), Vice-President of the Linnean Society, the famous 7th Superintendent of Calcutta Botanical Garden, and the major pivotal figure in Asian botany. In many respects Wallich was the foremost botanist of Asia, even to this day, yet no book about him has been published so far, though several detailed studies are currently in preparation (Arnold, in press., Sterll, in prep.). His life and botanical collections, correspondence and contributions were so formative that nothing less than a major collaboratory illustrated study would do him justice, though there are a great many smaller studies published in various journals. It is hardly possible here to give more than an outline of his life, emphasising his work relevant to Nepal, but it is hoped that some of the aspects explored are still sufficiently unknown to be of value, and that information and selected references can be made available to assist botanists and pteridologists in the Indian subcontinent in understanding and selecting types of Wallichian names. Wallich's life, work and herbarium-collections have been thoroughly documented by Smith & Parsons (1970) and de Candolle & Radcliffe-Smith (1981), the latter much drawn on here for dates and other information; some others of the most informative references to him are: Wallich (1824-1826;1828-1849; 1829-1832), Wight (1831), Hooker (1854), Bennett (1855), Thomson (1857), Christensen (1881, biography, 1924-1926, bibliography), Lee (1899), Anon. (1913), Burkill (1953: 867-876), van

Steenis-Kruseman & van Steenis (1950), van Steenis-Kruseman (1958), Sen (1962), Miller (1970), Smith & Parsons (1970), Mabberley (1980, Wallichian species), de Candolle (1981, *ined.*) de Candolle & Radcliffe-Smith (1981), Nayar & Das (1983), Stafleu & Cowan (1983), Desmond (1994), Shrestha & Press (2000, Wallichian Nepalese types). His portrait at the Linnean Society was published by Christensen (1924-1926) and de Candolle & Radcliffe-Smith (1981) and others; his letters are preserved at CAL (1832-1841!), K, G, UPS and a number of other archives of his correspondents. The Palmaceous genus *Wallichia* Roxb. was named after him as have been almost innumerable species of flowering plants and many ferns, with the epithets *wallichii* or *wallichiana etc.*

Nathan Wallich, as he was known to his friends, was born Nathan Wulff, in Copenhagen, Denmark, the son of a poor Jewish peddlar-merchant, Lazarus Wallich Wulff. In 1801 he went to study medicine and botany at Copenhagen University under Prof. Martin Vahl, a correspondent of Linnaeus, who worked on P. Forsskål's collections from the Yemen, and who was later succeeded by Prof. J.W. Hornemann. He achieved his MD and then went to India, arriving on 18th. Nov. 1807, appointed in advance as Assistant-Surgeon for the Danish settlement at Serampore (Srirampur), shortly north of Calcutta along the Hooghly river. But he was briefly arrested during the Danish-English War of 1808, part of the Napoleonic War, then was ordered to leave Calcutta to be an Army Surgeon at Birbhum, about 100 miles to the north-west (note by Wallich in Index to his Correspondence, by letter from Judge Robert Morrieson to Wallich, 4 Nov. 1814, CAL (!)) "Bierbhoom ...... where I had been appointed as Civil Surgeon, immediately on my nomination in the Compnys service; **friends** having exerted themselves to get rid of me at Calcutta. .... However the appointment never took effect, for being ordered to join the army then in the field, I was eventually put in charge of the Botanical Garden". Fortunately the Army was already away in the Field, so he was instead placed on assignment, according to his interests and abilities, to be an assistant to Dr. William Roxburgh at the Botanical Garden, then outside Calcutta, across the river, while from time to time continuing his Surgeon's work at Serampore. He thus began his spectacular botanical career, becoming under the employ of the East India

Company, for whom he eventually worked for another 31 years 7 months and 19 days (letter of Major W.M.M. Sturt, Military Dept., Govt. of India, Fort William, Calcutta, dated 3 April 1846, in Wallich correspondence, CAL (!), on his retirement on a pension of £300 per year). His friendly connection with Roxburgh was of considerable help to him in establishing himself as a botanist. In 1812 he made his first collecting excursion, going on voyage to Mauritius to recover from malaria. This visit enabled him to make many interesting botanical collections, which he afterwards wrote up and described in various papers, among his Indian collections. After qualifying as Assistant-Surgeon under the Bengal Government in 1814, he was appointed Acting Superintendent of the Botanical Garden in 1815, when Hamilton, who had succeeded Roxburgh at the Garden, retired early to Scotland due to illness. But being somewhat inexperienced and junior he stepped aside two years later for the appointment of Dr. James Hare, followed by Dr. Thomas Casey, until in 1817 he was himself most appropriately appointed as a most successful and dedicated Superintendent, on recommendation by Sir Joseph Banks and others.

During this first period of his career he made a number of important plant-collecting excursions. After Mauritius he collected in Bengal, including going near to the Nepalese border in Bihar in 1819, from where he hoped to be allowed to enter the country, but permission was not granted and his party was turned back, just as Hamilton's first near visit had been 20 years previously. Wallich was not able to visit Nepal until the next year, when he embarked upon his first major collecting expedition, arriving in Kathmandu on 21 Dec. 1820, where he stayed for a year before leaving the Valley on 8th November 1821 and reaching Patna in Bihar on 22 Nov. Permission for the visit was due to the overtures of his correspondent, the Resident, Edward Gardner, to the Royal Court of Nepal. Gardner had already been sending Wallich much material from Nepal (see above) and Wallich stayed at the Residency, where Gardner and his team, including the collectors Wallich had sent to him from Calcutta, had set up a botanical workshop for pressing plants, preparing seeds *etc.* in one of the Residency rooms. Unfortunately the political situation in Nepal at the time was only slightly improved from that during Hamilton's visit, as the Marquess of Hastings' friendly policy towards

British India's former enemy was only just beginning to melt the air of strong suspicion and bear fruit through Gardner's good offices. So Wallich was again confined to the Kathmandu Valley throughout his stay, apart from on the way up and on the return journey, even though (Wallich 1830: 36) he found he was generally less restricted than Hamilton had been. His collections from Gossainthan, north of the Valley, were made for him by Bharat Singh, one of his collectors from Calcutta, going on pilgrimage to the sacred lake, Gossainkund. Nevertheless he was able to collect a large amount of material on his way up as well as on the hills around the Valley and his localities (from his unpublished *Filicologia Nepalensis* in CAL (!), his *Tentamen Flora Nepalensis* (1824-1826), his *Numerical List etc.* (1828-1849) and his *Plantae Asiaticae Rariores* (1829-1832)) included Bichakori ("Bechiaco") to Dopabasa, south of the Churiaghat pass, on to the Rapti ("Rapty") Valley, up through Hetauda ("Hetaunra"), to Chisopani ("Cheesapany") mountain, "Ekdanta" Mountain and Bimphedi ("Bimpedy"), over to Chandragiri ("Chandaghiry") mountain, down to Thankot ("Tancote") to the S.W.; then in the Valley to Nagarjun ("Nagh-Arjun") mountain and Marekuh ("Marekoh/Marekow") village to the N.W., Sheopure ("Sheopori") mountain and Thoka ("Toka") spur near Thoka village ("Tokaganj") to the north, Swayembhunath ("Sumbhunanth") in the west-Valley, Pashupatinath ("Pusputinanth") and Raniban ("Ranibhund"), Gokarna ("Gokurrun") in the mid-Valley (N.E. of Kathmandu); and Sankhu ("Sankoo") to the north-east. While in Nepal, Wallich also despatched some of his Bengali collectors to the Gandaki valley in mid-west Nepal, west of Pokhara, from where a few collections were obtained and later sent replacement collectors up to Hodgson to continue sending a small number of selected plants and seeds.

Just as Hamilton had done, during his stay in Nepal Wallich wrote a detailed work on Nepalese Botany. This is an unpublished bound work in Wallich's handwriting on the ferns of Nepal, which had lain unnoticed among his papers for nearly 200 years in CAL (Fraser-Jenkins 2004), until unexpectedly rediscovered by the present author in July 2003 and transcribed by him in Dec. 2004. Until then Wallich's Nepal ferns had only been given the locality "Napalia" in his *Numerical List*, but many of them (apart from the last 20 species, which were either not written, or somehow not included in the bound volume) are now known with precise type- or syntype-localities. The work, called *Filicologia Nepalensis* (Wallich 1821, *ined.*) contains in clear handwriting, actually ready for publication, detailed descriptions in Latin of all his species of ferns, together with details in English of exactly where he collected them and how they are related to other species. Details of this book, which Wallich (1823) listed the contents of in July 1821 by number of species per genus, are given by Fraser-Jenkins (in prep.), with taxonomic and nomenclatural details of the species concerned. In addition it can be seen from the Index to Wallich's correspondence in CAL (!), that he wrote a great many letters to various botanists (including Gardner and others back in Calcutta and England) during the time he was in Nepal. Unfortunately these earlier Wallich letters (incoming and outgoing) are not in the 6 volumes now present in the two Archives cupboards in the new library at CAL, but as there were originally 31 volumes of his correspondence *etc.* given back to Calcutta by Kew in 1887 (van Steenis-Kruseman & van Steenis 1950, van Steenis-Kruseman 1958), they may still survive in some of the locked steel cupboards in the closed and abandoned old library, where a great many precious old volumes remain, in varying states. The present author could not find more letter-volumes there during a brief entry into the old library in Dec. 2004, neither could H.J. Noltie in Jan. 2005 (pers. comm.), but it is hoped that further search might reveal them as the keys to the cupboards are still extant. In fact the return of Wallich's letters and papers to Calcutta, though made for the best of purposes, subsequently turned out to be little short of a major disaster, which could hardly have been properly anticipated at the time. It has unfortunately resulted in their effective loss to Botany for over 150 years (apart from small parts of them cited by Prain, Gage *etc.*) and the fact that the chance for their survival is now extremely tenuous. Many, especially the continental European ones, written on more fragile paper, are falling apart at the edges, or worse, particularly at the tops of the bound volumes. It is very much to be hoped that they can be digitally photographed before it is too late. The whereabouts of the 25 volumes covering the earlier letters is at present a mystery as they are clearly listed in the Old Library catalogue. It is only to be hoped that they have survived in the locked cupboards and are still capable of being read. The whole series is of intense Botanical and Historical interest.

As already explained, Wallich sent Gardner's Nepalese collections back to Britain (to Lambert and Smith) prior to his going back there himself, but his own, more extensive collections from 1820-1821 were brought back to Calcutta by him and later taken to England in 1828, so were not worked on by Don. Wallich's own collections from Nepal thus became part of the numbered series listed in his Numerical List and are thus quite a different category from Gardner's earlier sendings, even though both are given as from Wallich and Gardner's name is not mentioned on the sheets of the earlier collection. A number of specimens from his 1820-1821 collection were given names by him on the sheets that do not appear in his *Numerical List*; these are specimens that he sent back to correspondents in England between 1821 and his own return in1828, and the names have been cited as "Wall. in herb." In a few cases he also gave names to some of Gardner's earlier collections, which have also been cited as "in herb.", but unless the earlier date is given on the specimen, most "in herb." names are from the later collections. It might be possible that a few of his early collections could also have been included by him in the later numbered and catalogued sets, but this can only apply to a very small number of specimens, if at all, as nearly all the early material was no longer in his hands when he worked on the cataloguing after 1828. If specimens from the later Wallich collections at K.-W or in K (labelled "Herbarium Hookerianum 1867") must be used in typification of a Don name due to the absence of any suitable early Wallich material, strictly these are only neotypes, and not lectotypes.

At the end of his Nepal visit he unfortunately went down again with malarial fever in the plains on the way back to Patna, and became seriously ill for the first part of 1822, having to take medical leave for a voyage to Malaya (Malacca and Penang) and Singapore. His health recovered and he was soon making the best of the opportunity to go collecting again. He also became acquainted with Sir Stamford Raffles, the founder of the new territory of Singapore, who gave him permission to start a new Botanic Garden there, now the famous Singapore Botanic Garden, and also granted him an estate on the island which he named Dannebrog Hill, but which was later known as Mount Wallich (though now levelled for landfill purposes). A further meeting was with his Malaysian wife, to whom he remained happily married all his life and by whom he had two sons, Charles and David and a daughter, Hannah. On his return to Calcutta at the end of 1822 Wallich completed the publishing of Carey's edition of Roxburgh's *Flora Indica* (1820-1824), though without Roxburgh's ferns, which were not published until 20 years later, edited by Griffith (1844). The original manuscript of Roxburgh's in CAL (!), though now in poor state, still survives in a small packet in the new library and is useful in showing the changes added to it by Wallich and Griffith. Wallich also published his *Tentamen Floræ Nepalensis Illustratæ* (1824-1826), and made two further major collecting expeditions, the first to Awadh ("Oudh"), in what is now west-central Uttar Pradesh, visiting Lucknow, Saharanpur, Haridwar, Dehra Dun, Meerut, Delhi and Kanpur, in 1825. His second expedition was to Myanma (Burma) and the Court of Ava (at what is now Mandalay), on John Crawfurd's diplomatic mission, again following partly in the footsteps of Hamilton, in 1826 and 1827, but collecting more herbarium material, returning via Moulmein, Amherst, Chappedang river Martaban and the Saluin river, and bringing the extensive material back to Calcutta.

The next phase of his life was a new burst of energy stimulated by his return to Britain in 1828 on sick-leave, with his wife and family, which he managed with considerable difficulty to extend while there to the Company's maximum allowed five years. He obtained permission to take with him virtually the whole of the herbarium in order to work on it and catalogue it in London. It was a measure of his foresight that he realised he could not hope to complete this Herculean task himself in Calcutta, and unknown to him, it is undoubtedly only because of this that the collections of 19th century botanists in India still survive today. Due to neglect and the ravages of insects and the difficult climate, nearly the whole of the original "good", large and fairly complete Wallich set sorted out for Calcutta by Sir Joseph Hooker and Dr. T. Thomson and brought back to be given to the Botanical Garden by Thomson in 1855 had been destroyed by 1960, though they were still present in the late 19th Century. In all some 11,000 Wallichian sheets, were received by Calcutta, the main part brought there by Thomson on his return from Kew to India, while extra collections were sent out under Kew distribution numbers up to 1869 (Thomson (1857), King (1899), Anon. (1912: 5), Anon. (1913), Hill (1938), Chatterjee (1948), Burkill (1953: 876), van Steenis-

Kruseman (1958), de Candolle & Radcliffe-Smith (1981)). In fact their very existence has even been denied by some later workers (Ghosh 1947-1948, Nayar & Das 1963) and it has become an oft-repeated nationalist Independence-myth in India that the colonial British were so exploitatory as not even to give a Wallich set to Calcutta. This theory was put forward in error of the true facts by Ghosh (1947-1948), who did not cite fully what was recorded, and by occasional Indian botanical visitors to Kew, even in recent times, as a reason to ask for the return of all Indian specimens from Kew *etc.* to Calcutta! Had that happened there is no reason to suppose their fate would be any different from the loss of a great number of older sheets there, including even the Nepal collections of Stainton, Sykes and Williams in 1954, and a number of Panigrahi's Assamese specimens of the 1960s, which are vitally important given the wide misidentifications he published, many now having been very badly eaten by insects due to intermittent failure to place moth-balls in the cupboards. Many of the old and even recent sheets crumble into flakes on being handled, while the majority reveal moderate to severe insect-damage, the odd specimen not infrequently being little more than an impression in the glue with some remnant fragments (even though these are often still identifiable and useable). In 1991, the present author actually purchased and placed mothballs himself at his own initiative in the fern-herbarium, when the cupboards were in danger of becoming more similar to an entomological menagerie. In the 1980s he saw a number of old Wallichian and other sheets that must have been survivors of the original set, but had frequently had their labels cut in half with only the tops of words still visible, or worse, in order to cut the sheets down to fit them into the new, smaller cupboards, which had obviously been done by *paeons* with no knowledge of how important the label is! As a result of these problems a replacement set of duplicates was requested from Geneva in 1958-9 (Nayar & Das 1963/1983 BBSI), which mostly survive now, though some of these are also in perilous state in 2004.

Wallich arrived in London in July 1828 and in addition to the vast herbarium-collection in 30 great barrels, weighing some 20 tons, had 52 chests and 12 cases of living plants (when he returned to Calcutta in 1832 he pioneered the use of Wardian cases, letter of N.B. Ward to Wallich, 11 July 1834, CAL (!)) as well as a surviving pair of Tibetan Dogs from

Nepal, sent by B.H. Hodgson, to unload off the Ship, "The Orient", at Tilbury Dock (Wallich to P. Auber, EIC, 16 July 1828, CAL (!)). On 18th Sept. 1828 (letter at CAL!) the Honourable Court of Directors of the East India Company agreed to his plan to work on the material and publish it, "on the clear understanding that the Company are not to incur any additional expense whatsoever on this account [publishing]". They initially allowed him two and a half year's leave, but would not give any additional allowance, beyond his normal 1000/= rupees per month to cover his household, the 3 assistants or servants he had and all the expenses of preparing and distributing the specimens. He thus found premises to rent at 61, Frith Street, Soho, London, and converted them into a very crowded museum-workshop to sort the specimens into many sets, and catalogue and number them in his *Numerical List etc.* (1828-1849), which he did with the help of a variable team of botanists, including Lindley, George Bentham, A.L.P. de Candolle, of Geneva and Prof. R. Graham, of Edinburgh. Graham later took charge of Wallich's son, George Charles Wallich (1816-1899), as a student at Edinburgh, after Wallich's return to Calcutta, though George "became excessively idle" and fell into debt in bad company until a Sherriff's warrant was issued for his arrest and commitment to prison, Graham paying the bill of nearly £700 - which must have straightened him up somewhat in preparation for his distinguished future oceanological career! (Graham to Wallich 26 Feb. 1835, CAL (!)). The work of distribution of the material steadily went ahead throughout his 5 years leave, up to no. 7683, the later numbers, up to 9148, being completed after Wallich's final retirement to England in 1846. After 1829, the E.I.C. also allowed Wallich to catalogue and distribute various other herbaria already lying in their Indian Museum, including those of the south Indian Moravian missionaries, of Hamilton from the Bengal Survey, some of Roxburgh's remnants and the big herbarium of Robert Wight, from south India. As a result of the great ongoing industry at Frith Street, Gardner, then retired to London, wrote to Wallich on 4 May 1832, "My Dear Wallich, It is not an easy thing to catch you – but I shall call at your "Kar-Khannah" [Nepalese for "factory"] one of these days, where I shall most likely find you. I have a letter from Hodgson in Nepal, late in Decr. – who, among other things, asks after you ....". Wallich's most appropriate and altruistic purpose was not only to catalogue the specimens, under new or well known names, but also to spread them out to every centre where Botanical study was taking place, with a view to the Indian flora becoming known to science.

Details of the number of collections in the herbarium were given by Smith & Parsons (1970) and de Candolle & Radcliffe-Smith (1981), the collectors or collections relevant to Indian ferns were: collections from Calcutta Botanic Garden, labelled H.B.C.; Hamilton's Bengal *etc.* plants; some of William Roxburgh's S. Indian and east Indian-peninsular collections; the Tranquebar Missionaries, Röttler, Klein and Heyne, from S. India; Robert Wight's S. Indian plants; J.G. König, the founder of the Tranquebar Missionaries, from S. India; Robert Blinkworth, a collector employed by Wallich in Kumaon and Garhwal (Uttaranchal), who visited Srinagar (Garhwal), Badrinath and Kedarnath, and also helped label Wallich's collections in large and distinctively square-lettered writing; Kamrup, a collector employed by Wallich in Sirmoor, Srinagar (Garhwal) and Badrinath (Kumaon, Uttaranchal), a few of whose collections had been available to Don and are included in his *Prodromus* as Kumaon was formerly in the possession of Nepal, not to be confused with the District of Kamrup in Assam, just south of Bhutan; Francis de Silva, a collector employed by Wallich in Sylhet (Bangladesh), who collected from a boat at Pundua (Bangladesh) in 1821, and went up to the Khasi Hills (Meghalaya), having previously collected in Nepal in 1817-1818; M.R. Smith, Magistrate with a garden at Sylhet (Bangladesh), who collected up to the lower Khasi Hills (Meghalaya), and had also collected for William Roxburgh, Wallich's predecessor at Calcutta; Henry Bruce, a collector employed by Wallich in Sylhet and Chittagong (Bangladesh); Dr. J.F. Royle, second Superintendent of Saharanpur Garden (Himachal Pradesh) for the E.I.C. and collected in Himachal Pradesh (including Mt. Choor Chand and Sirmoor); Sir Robert Colquhoun, British Resident in Kumaon in 1819 and in charge of the Kumaon Gurkha Battalion, who collected in Kumaon (Uttaranchal) and visited Gardner to collect in Kathmandu in 1818; Major-General Thomas Hardwicke, who collected in Srinagar, Garhwal and then Himachal Pradesh and Kashmir in 1799 and supervised the making of a series of Indian paintings, now in BM, including ferns; William Moorcroft, the explorer, collected briefly on a visit to Kathmandu Valley in 1820, but extensively from Himachal Pradesh, Ladakh, N. Pakistan and finally Afghanistan, whence, he died of fever on the way back from Bokhara; Dr. George Govan, the first Superintendent of Saharanpur Garden, who collected near Simla (Himachal Pradesh); Capt. W.S. Webb, surveyor, who collected in Kumaon and worked with Blinkworth and Kamrup; the Gerard brothers, including Capt. Patrick Gerard, who collected in Sirmoor (Himachal Pradesh and Uttaranchal).

During this most active time in England, Wallich made a great many botanical and horticultural friends, though nowadays we might be forgiven for thinking he was rather obviously cultivating the rich and famous, which was much of a necessity in those days. His letters and their Index in CAL reveal that if a well known, or titled aristocrat had an interest in horticulture of Asian plants at his park, it was quite likely that Wallich would sooner-or-later be making his best efforts to visit and supply rare Indian orchids, or seeds *etc*., and he also appears to have skillfully increased his income thereby. It was while in England that he also prepared his magnificent *crême-de-la-crême* of Botanical works, his lavishly illustrated *Plantæ Asiaticæ Rariores* (1829-1832), one of the finest productions of the Century, in which many Nepalese species, including a couple of ferns were described. The black-and-white Indian reprint of this work, though useful, cannot begin to do it proper justice compared to the pleasure of opening and admiring the breathtaking pages of the original (BM!). It is interesting to find that in 1832, he had planned and sent out a prospectus (now in the Indian Office of the British Library) for a companion two-volume set on the ferns, *Filices Asiaticæ Rariores*, to be produced by Hooker & Greville at Glasgow and Edinburgh, but the E.I.C. refused to subscribe to more than 6 copies and the work was shelved for financial reasons, most or all of the ferns later being published and illustrated by Hooker, or by Hooker & Greville, in their well known works on ferns.

It was Wallich's remarkably well executed plan to distribute the herbarium to any who would work thoroughly on it that brought the greatest benefit to Botanical science. A list of those who worked on various flowering plant genera has been given by de Candolle & Radcliffe-Smith, and a further list of where Wallich sent material is as follows, from Wallich's handwritten lists (Wallich to P. Auber, E.I.C., 20 June 1831, 14 March 1832 and 30 Sept. 1832, CAL (!)):

1. Prof. Agardh, Lund in Sweden (6 bundles), now in LD.
2. Mr. G.A.W. Arnott, Edinburgh (32 bundles), now in E.
3. Mr. Battley, London (3 bundles).
4. Mr. Geo. Bentham, London (16 bundles), now in K.
5. The University of Berlin (10 bundles) + 8 palms, now in B (destroyed in 1943, except ferns).
6. Dr. C. Blume, Brussels (8 bundles), now in L.
7. Baron Bory de St. Vincent, Paris (1 bundle), now in P.
8. The British Museum (29 bundles) + 18 palms, now in BM.
9. M. Adolphe Brogniart, Paris (6 bundles), now in P.
10. Mr. Brown, London (35 bundles) + 5 palms, now in BM, K, E.
11. M. Cambesedes, Paris (2 bundles), now in MPU.
12. Profr. P. De Candolle, Geneva (34 bundles), now in G.
13. Profr. A. De Candolle Jr., Geneva (9 bundles), now in G.
14. Baron Benjamin Delessert, Paris (27 bundles), now in G and G-DC.
15. M. Duvan, Paris (1 bundle).
16. M. S. Endlicher, Vienna (3 bundles), now in W.
17. Profr. Fischer, St. Petersburgh (8 bundles), now perhaps in B (destroyed in 1943, except ferns), BR, G, K, KIEL, OXF, P, or W (see Lanjouw & Stafleu 1957).
18. Profr. Graham, Edinburgh (48 bundles), now in E.
19. Dr. Greville, Edinburgh (24 bundles), now in E, BM, GL.
20. Mr. A. Haworth, Chelsea (4 bundles), now in OXF, K.

21. Profr. Henslow [misspelt in one list, which was not in Wallich's writing, unlike the others, as "Henclaw"], Cambridge (11 bundles), now in CGE, K.

22. Profr. Hooker, Glasgow (32 bundles), now in K.

23. Profr. Hornemann, Copenhagen (8 bundles), now in C.

24. Profr. Adrian de Jussieu, Paris (11 bundles), now in P-JU.

25. Museum of King's College, London (4 bundles) + 4 palms.

26. Profr. Kunth, Berlin (25 bundles), now in B (destroyed in 1943, except ferns).

27. Mr. A.B. Lambert, London (20 bundles), now in BM, CGE, OXF, B (destroyed in 1943, except ferns), FI-W.

28. Profr. Lehmann, Hamburg (12 bundles), now probably in K, HBG, S.

29. Profr. Lindley, London (24 bundles), now in CGE.

30. Linnean Society, London (17 bundles) + 3 palms, now in LINN.

    [This was before the major donation received by Francis Boot, M.D., Secretary of the Linnean Society, on 29-30 Sept. 1832 (letter in CAL!), of 55 bundles and 25 boxes, the whole remnant of dried plants under Wallich's charge from the E.I.C.].

31. Museum of London University (4 bundles) + 12 palms.

32. Profr. von Martius, Munich (19 bundles), now in BR, MEL.

33. Profr. Meissner, Bale (12 bundles).

34. Museum at Prague (via Count Sternberg) (5 bundles) + 5 palms, now in PR.

35. Mr. Arch. Menzies, London (3 bundles), E.

36. Profr. C.G. Nees von Esenbeck, Breslaw, now in BONN, CGE, GZU, STR *etc.*

37. Profr. Fred. Nees von Esenbeck, Bonn/Breslaw (20 bundles) + 3 palms, now in BONN, CGE, GZU, STR *etc.*

38. Museum d'Histoire Naturelle, Paris (6 bundles) + 5 palms, now in P.

39. Mr. John Prescott, St. Petersburgh, now at OXF.

40. Public Museum, Christiania, Norway (3 bundles) + 5 palms, now in O.

41. Public Museum, Florence (5 bundles) + 5 palms, now in FI.

42. Public Museum, Munich (4 bundles) + 5 palms, now in M.

43. Public Museum, Turin (5 bundles) + 3 palms, now in TO.

44. Profr. Reinwardt, Leyden (5 bundles) + 3 palms, now in L.

45. Profr. Achille Richard, Paris (11 bundles) + 3 palms, now in P.

46. Dr. Roeper, Bale (5 bundles) + 3 palms, now in P, ROST.

47. Profr. Jul. H. Schultes, Landshuth in Bavaria (11 bundles) + 3 palms, now in M.

48. Dr. Schweinitz, Pensylvania (7 bundles) + 3 palms, now in PH.

49. Mr. Shepherd, Liverpool (1 bundles) + 3 palms, now in LIV, OXF.

50. Mr. Smith, Kew Garden (1 bundles) + 3 palms, now in BM.

51. Zoological Society, London (17 bundles) + 8 palms.

52. Society of Apothecaries, London (2 bundles) + 8 palms.

53. Profr. K. Sprengel, Halle (13 bundles) + 8 palms, now perhaps in B (destroyed in 1943, except ferns), his ferns in LZ were destroyed in 1943, parts of his herbarium survive in various other herbaria (see Vegter 1986).

54. Mr. C. Stokes, London (3 bundles) + 8 palms.

55. Profr. A.T. Thomson, London (2 bundles) + 8 palms, now ?in CRK.

56. Dr. Trinius, St. Petersburgh (7 bundles) + 8 palms, now in LE.

57. University of Berlin (10 bundles) + 8 palms.

58. University of Cambridge (10 bundles) + 7 palms, now in CGE.

59. University of Edinburgh (3 bundles) + 8 palms, now in E.

60. University of Glasgow (3 bundles) + 7 palms, now in GL.

61. University of Oxford (3 bundles) + 6 palms, now in OXF.

62. University of Vienna (11 bundles) + 6 palms, now in WU.

63. Imperial Academy at Vienna, now in W.

64. Profr. Wahlberg, Stockholm (5 bundles) + 6 palms, now in S, UPS.

65. Mr. Ward, London (2 bundles) + 6 palms, now in BM, K, OXF.

66. Profr. Williams, Oxford (6 bundles) + 6 palms, now at OXF.

Most recipients were also given 5 tree-trunks (mostly of palms), unless otherwise mentioned. Apart from those above and the early Wallich collections mentioned previously, other sets of Wallich material, are in DBN, PRC, MICH, AMD and RO.

Wallich (*in litt.* 30/9/1832, and 30/2/1832, CAL (!)) said that altogether 641 parcels of herbarium-specimens (some 226, 000 duplicates) and 83 sections of palms or other woods had been sent out, and in addition more material was distributed after 1846. This enormous service to the scientific knowledge of mankind was aptly appreciated by de Candolle (1981, *ined.*), from letters of his great-grandfather, A.L.P. de Candolle, among others. But all too soon, after having been granted a further extension of leave, the time came when the E.I.C. could not grant any more without breaking the strict provision for officers to have an absolute maximum of 5 years while remaining in the Company's employ. Wallich made unsuccessful attempts to extend it further in order to complete the cataloguing and distribution, but in the end was almost press-ganged by the E.I.C. on to his ship, "The Exmouth", at Tilbury, and sailed on 10 Oct. 1832. He had been sent off with a blaze of "Last supper" dinner-parties, including one with Roxburgh's widow and family, and salutations from his many friends, some of whom, like Wight, felt that his return to pestilential Calcutta might mean that they never met again. But the E.I.C. stipulated that his salary backlog over the last five years would only be paid to his agents, *in Calcutta*, to be handed over on his arrival. Not only that, but they would not pay his passage money, and insisted that he paid them back an advance of 2212 rupees he had been given in Jan. 1828 to tide him over on arrival in Britain, even demanding that he pay interest on it of 8% per annum, if his salary were to continue! It seems sometimes that some of the egregious acknowledgements in his dedications of books to the generous munificence of the EIC, were made precisely because the opposite was really the case on most occasions! Prior to his departure he had an interesting meeting with Prof. Charles Babbage, who came to show him his "Great Engine", or Difference Engine to 15 figures, which "works perfectly" (Babbage to Wallich 6 Oct. 1832, CAL!), the first calculator machine. This was no pocket job as it was about the size of a piano, but unfortunately it broke down and the embarrassed Babbage regretted "that a small part of the Engine did not sooner exert its cogitating powers that you might have been a witness for me on the other side of the Globe"! Before Wallich left England, he made crucial final arrangements for the Indian Museum to hand over the complete remnants of their herbarium material and all he had at Frith Street (including his top, best set), to the Linnean Society, together with copies of the Catalogues of Hamilton's Bengal herbarium and Heyne's and Röttler's Herbaria. It was this move that eventually founded the Kew Wallich herbarium (K-W), since the top set was later given by the Linnean Society to Kew in 1913, where it now remains. He also obtained the Linnean Society's undertaking to separate a set to be sent back for his use in Calcutta since he had been unable to due to the overwhelming amount of work he had to do before leaving, but due to lack of staff and their preoccupation with other major work (the ongoing work of Royle at the Linnean Society, see Burkill 1953: 876), this was not done until Hooker and Thomson sorted out a good set for both Kew (now found in the general herbarium without the "Herbarium Hookerianum" stamp), who did not have one, as Hooker's private set was not given to them until 1867, and Calcutta, which Thomson brought back with him to

Calcutta in 1855, thus fulfilling Wallich's request and the Linnean Society's promise, but some 20 years later.

When Wallich got back to Calcutta in early 1833 it was to find that the E.I.C. had instituted a major cut back in funds, with his old collectors laid off and looking to him for help he could hardly give (though from his generous nature, he did manage to help some personally) and major trips curtailed. The third phase of his life was therefore one of enforced reduction in his activities, also dictated by his returned and continuing illness. He spent much of his time sending plants from the Garden to well known gardens in Britain and Europe. But one major excursion arose, the Assam Tea Delegation of 1835-1836, to go into upper Assam and investigate the occurrence there of tea, whose presence had been confirmed by Capt., later Colonel Francis Jenkins, the surveyor and political officer in Assam, the year before (though his rôle in its discovery was nearly suppressed by others, see letter of Rev. David Jenkins, of Mevagissey, brother of Francis, to Wallich, of 28 Dec. 1838, CAL (!)). Their study was intended to investigate under what conditions tea was growing and to bring back plants with a view to setting up the planting of tea as a commercial product for the E.I.C. Their route led up the Brahmaputra into the virtually unknown, wild tribal lands in what is now N.E. Arunachal Pradesh, towards the north Myanma (Burmese) border - some of the richest forest areas in the Indian subcontinent, which are still poorly known and difficult of access today. However the trip came to grief as Wallich's companions, the botanist, Dr. William Griffith (1810-1845), who died prematurely in Malacca, after editing Roxburgh's ferns (Griffith 1844), and Dr. John M'Clelland, Geologist, were unable to get on with the older Wallich, who wanted to take control of what they were to collect and could not fit in with them as a team. Their personality clashes, a common problem on expeditions to remote places, could not be resolved and Wallich, who was also unfit from his illness, left the party in lower Assam, while the others went on, Griffith going up into north Burma for the next two years, making many spectacular collections, now at Kew, BM and elsewhere.

After a while, Wallich became more seriously ill back in Calcutta and was obliged to take a voyage to Capetown, S. Africa, for two years in 1842, returning to find his old adversaries of the Tea delegation had damaged his beloved Botanic Garden with new planting-schemes of their own and had had the temerity to submit damaging and unfair reports against him (Griffith 1843, M'Clelland 1841, 1848, Burkill 1956). The subsequent temporary appointment of M'Clelland in the garden marked the sad end to Wallich's stay at the Botanic Garden he had done so much to create. He no longer had the will to stay on and within a year attempted to resign, but was told that resignation was not possible, except on health grounds. It rather appears from his letters in CAL (!) that he then over-emphasised his generally poor health, and though his colleagues did not expect him to succeed in a "touch and go case" they were pleasantly surprised on his behalf that he his resignation was accepted, offering congratulations on his "escape ..... from this land of Cholera", with pension intact (D.D. Bourne, Bengal Govt., to Wallich, 4 April 1846, CAL (!)). His retirement was much helped by his wife's having inherited a large legacy of £20,000 pounds (the equivalent of about a million pounds today) from her Aunt (letter from Wallich to Wight at Kew (Kew Director's Correspondence 53/153), dated 17 Nov. 1839, kindly provided by Dr. H.J. Noltie).

Wallich sailed back on the "Hindoostan" and began a much more pleasant and highly deserved fourth stage of his life back in England. This was probably the most enjoyable phase, once he had taken a holiday with his family by the beaches in Normandy to completely recover his health. He lived at 5, Upper Gower Street, London, and was an active Vice-President of the Linnean Society and fellow of the Royal Society and the many other Societies he was a member of. He was also able to make tours visiting botanical and horticultural friends in their great gardens in Britain, advising and bringing seeds *etc.* The letter from his old friend, Lady Sarah Williams, formerly Amherst, wife of Sir John Hay Williams, Bôdchwyddan, St. Asaph, in April 1847 (CAL!), is rather typical of his enjoyable socio-horticultural existence at this time – "Dear Dr. Wallich, I do not know if you will recognise me under my present name as a member of the Amherstia family, & an old acquaintance -- Sir John Hay Williams & myself would be much delighted to receive you in Wales & welcome your return to England. — He has a choice and thriving collection of Stove plants. — Approachable by railhead via Liverpool or Chester." On 15 May 1847 Sir John wrote, sending a pest from his greenhouse – "The horrid creature has appeared again in my Exotic House & I send you a specimen of it – pray tell me what remedy would be the best to get rid of it." In 1847 he joined hands with Bentham to complete the last 1500 numbers of the catalogue and herbarium distribution of his *Numerical List*, making many visits to Mr. and Mrs. Bentham's Pontrilas House, near Hereford, where they worked together to complete the huge labour, much to everyone's satisfaction. He also attended with them the meeting of the Swansea Association and Royal Institution, staying with the Dillwyn-Llwellyns at Penllergare House, near Swansea, and visiting the Trahernes at Coedarhydyglen ("Coedriglan") House, near Cardiff (both families the present author used to visit as a child at their same Glamorgan County houses) in June 1848. His last years were occupied in successfully completing and publishing a number of papers from his Indian manuscripts.

On 28 April 1854, the great man, who had risen by his remarkable energy and application, died at the age of 68 at his London House and was buried, like Don in the cemetery at Kensal Green, London, leaving behind him an enormous botanical legacy. Studying his surviving letters in Calcutta it is easy to see that he was a most kind-hearted, warm and positive personality, not only from his outgoing letters, but in many letters from those around him, both European and Indian, and further afield, who clearly held him in a respect borne of affection for his caring and humanitarian nature. His enemies were few, though they clearly included Griffith and M'Clelland, both of whom appear to have been less than generous, bearing in mind his greater age and experience, the fact that he had personally created the successful structure of the enormous botanical set up concerning Calcutta and beyond, and that he was constantly suffering from fever and illness, which would have sapped many a younger man's energy. In addition he was an obviously Jewish foreigner from Denmark, with a Malay wife, disadvantages indeed in the exclusivist atmosphere of British Society in India, yet managed to become endeared to nearly all around him, with no bitterness. The touching letters to him from the Revd. Bernhard Schmid, a Moravian missionary and botanist at Ootacamund, South India, at the time when one of Wallich's infants died of fever (Schmid had lost all his children and his wife in the space of a few months), show that he was a deeply yet quietly religious man, accepting the spiritual value of Christianity, without being carried to unbalanced extremes. One measure of the general admiration for him was the number of Honours and Memberships of Societies he had, though he obviously did set out to cultivate Societies for the connections they afforded, unlike the spirited B.H. Hodgson, who in a letter back to him, of 20 June 1833, after a tiff with the Zoological Society, completely disdained " those who ....cringe and crave in order to ... write sundry capitals after their names..... I never asked, nor ever will, to be made a fellow of any Society ..... I hate & abominate the whole pack of these assorted gents!"

As recorded in the Index to Wallich's letters at Calcutta, up to 1832, his honours were as follows:

1. 25 April 1806 - Diploma of the Royal Academy of Surgeons, Copenhagen, having successfully passed the examination as an Assistant Surgeon.

2. 6 Sept. 1806 - Requisition from the Danish Govt. at Fredericksnagar (Serampore), Bengal, via the Royal Board of Economy and Commerce, Copenhagen, to the Royal Board of Health, Bengal Establishment, for a Surgeon.

3. 26 March 1816 - Appointed as Surgeon to the Salt Agency of the Twenty-four Pergunnahs [a district now including the surrounds of Calcutta], without prejudice to his being Assistant to the Superintendant of the Botanic Garden, Calcutta.

4. 1 March 1817 - Appointed as Acting Superintendant of the Botanic Garden, Calcutta, promoted by Sir Joseph Banks.

5. 18 Nov. 1817 - Commissioned as Assistant-Surgeon in the East India Company's Bengal Establishment, with effect from 10th May 1814.

6. 17 Nov. 1817 - Certificate as a Candidate for Admission into the University of Copenhagen, as Assistant Surgeon.

7. 6 Jan. 1818 - Diploma of appointment as a Corresponding Member of the Horticultural Society of London.

8. 8 June 1818 - Warrant as a Knight of the Royal Danish Order of Dannebrog [St. Danneburg], conferred on him by His Majesty the King of Denmark.

9. 10 June 1818 - Commissioned as Assistant-Surgeon, with that rank in the King's Army - signed by the Marquess of Hastings, with effect from 10th May 1814.

10. 23 Dec. 1818 - Appointed a Corresponding Member of the Society of Arts, London.

11. 25 Dec. 1818 - Diploma as a Member of thc Veterinary Society of Copenhagen.

12. 7 Feb. 1819 - Election as a Fellow of the Linnean Society of London.

13. 20 March 1819 - Election as a Member of the Geological Society of London.

14. 30 April 1819 - Diploma of M.D. at the University of Aberdeen, Scotland.

15. 17 July 1819 - Election as an Honorary Member of the Library Society of Madras.

16. 26 Oct. 1819 - Election as a Corresponding Member of the Academy of Natural Sciences of Philadelphia, USA.

17. 1 Jan. 1820 - Diploma as a Member of the Academia Caesarea Leopoldino-Carolina Naturae Curiosorum, Breslaw.

18. 19 July 1820 - Election as a Foreign Corresponding Member of the Société Royale d'Agriculture, Paris.

19. 30 Sept. 1820 - Diploma of appointment as an Honorary Member of Liverpool Botanic Garden.

20. 2 Jan. 1821 - Diploma as a Corresponding Member of the Musée Royale d'Histoire Naturelle, Paris.

21. 2 Jan. 1821 - Diploma as a Doctor of Philosophy in the Royal University of Copenhagen.

22. 11 March 1822 - Diploma as a Member of the Societas Caesarea Naturo Curiosorum Mosquovis, Moscow.

23. 2 Dec. 1822 - Election as an Ordinary Member of the Royal Society of Edinburgh.

24. 1 Jan. 1823 - Election as an Honorary Member of the Association for the Encouragement of Horticulture in Prussia.

25. 1 Jan. 1823 - Election as a Corresponding Member of the Economical Society of the Philippine Islands.

26. 20 Feb. 1823 - Grant of over 19 Acres of Land at Tillock Ayer, Singapore, to be known as Dannebrog Hill, approved by Sir T. Stamford Raffles, Governor of Singapore.

27. 24 May 1823 - Diploma of the Honorable Association of Berlin Botanic Garden.

28. 12 June 1823 - Election as Corresponding Member of the Batavian Society of Science and Arts.

29. 2 July 1823 - Diploma of Carrera-Mantia of the Society of Science and Arts, Batavia.

30. 2 Jan. 1824 - Diploma as a Corresponding Member of the Museum d'Histoire Naturelle, Paris.

31. 29 April 1824 - Election as Honorary Member of the Philosophical and Literary Society of the Bristol Institution.

32. 15 May 1824 - Election as an Honorary Member of the Agricultural Society of St. Helena.

33. 30 May 1825 - Appointed a Fellow of the Geological Society of London.

34. 1 Nov. 1825 - Diploma as a Foreign Member of the Physiographical Society of Lund, Sweden.

35. 2 Nov. 1825 - Election as an Honorary Member of the Royal Botanic Institution of Glasgow, Scotland.

36. 12 May 1826 - Commission as Surgeon in the service of the Hon. East India Company, as from 5 May, signed by Earl Amherst, Governor General.

37. 4 Nov. 1826 - Election as a Member of the Royal Asiatic Society of Great Britain.

38. 15 Dec. 1826 - Diploma as a Member of the Royal Society of Sciences, Copenhagen, dated 30 Dec. 1823.

39. 23 June 1827 - Commission as Full Surgeon with Corresponding Rank in the Hon. East India Company's service, as from 5 May 1826.

40. 13 March 1829 - Election as Fellow of the Royal Society, London, as from 12 March.

41. 9 Aug. 1829 - Warrant of the Silver Cross of Dannebrogsmand, of the Royal Danish Order of Dannebrog.

42. 20 March 1830 - Diploma as a Member of the Royal Botanical Society at Regensburg.

43. 31 May 1830 - Election as a Corresponding Member of the Royal Academy of Sciences of the Institute of France, signed by Baron Cuvier.

44. 6 July 1830 - Diploma as a Foreign Member of the Horticultural society of London.

45. 7 Aug. 1830 - His Majesty King William IV's command to insert his name in the list of subscribers to *Plantæ Asiaticæ Rariores.*

46. 9 April 1831 - Diploma as an Honorary Member of the Museum of Bohemia, Prague.

47. May 1831 - Election as an Honorary Corresponding Member of the Aberdeenshire Horticultural Society.

48. 20 Nov. 1831 - Diploma from the Musée d'Histoire Naturelle, Paris.

49. 1 March 1832 - Elected as a Corresponding Member of the Zoological Society of London.

    Wallich listed the following additional Honours on the title-page of *Plantæ Asiaticæ Rariores* **1** (1829-1832):

50. Member of the Academy of Natural Curiosities of Bonn.

51. Member of the Asiatic Society of Calcutta.

52. Member of the Agricultural Society of Calcutta (he was actually the founder-member of the Agricultural and Horticultural Society of Calcutta).

53. Member of the Medical Society of Calcutta.

54. Honorary Member of the Helvetic Society of Natural Science.

55. Honorary member of the Royal Society of Ratisbon.

His plan to ensure the nearly impossibly large task of naming and documentation of the vast British-Indian collections by apportioning them, family by family to all the specialists of his time was ultimately highly successful, and perhaps the only strategy that could have succeeded. One of the quirks of this plan is that because he merely gave bare names, or *nomina nuda*, in his *Numerical List* the bulk of Wallichian species were not validly described by him (even though the names used to be accepted in Victorian times as valid, giving just the citation "*Wall.*"). Although his combinations were valid, as were any of the species he gave a description to in his other books and published papers, a very few being in the *List*, the species must be attributed to the author who actually validated the Wallichian name. Thus the citation should omit "Wall." and just put "Mett." if the genus was changed from what Wallich made it by the validating author, or it should be "Wall. *ex* Mett.", if the name is fully identical to Wallich's and was attributed by an author to him by reference to him.

This also means that the selection of lectotypes should best be made from the herbarium the particular author concerned studied, not necessarily from the K-W collection, and almost always not from the fragmentary and often poorly labelled survivors of the original and later Wallich sets in CAL. Many workers have too easily assumed that the holotype or lectotype of a Wallichian name, and also those of Don's names, are usually to be found in K-W, but in more cases than not, this is not so. With reference to Wallich's ferns, and particularly those from Nepal, the main authors who validated his names are (abbreviated according to Pichi Sermolli (1996) and with main herbarium in order of importance for selecting their Wallichian types): D.Don (BM), J.Agardh (LD, K-W), C.Presl (PRC, PR), T.Moore (K), R.Sim (based on Moore, K, K-W), J. Sm. (K, BM), Spring (BR, LG), Griff. (K-W), Kunze (B, K [Kunze's own herbarium is at [LZ, completely destroyed in 1943], JE, K, B, TUB]), Mett. (B), Kuhn (B), Fée (K, P, MPU, RB, STR), Ettingsh. (B), Hook. & Grev. (K, E), Grev. & Hook. (K, E), Hook. (K), Hook. & Arn. (K, E), Hook. & Baker (K, K-W), Baker (K, K-W), Bedd. (K, BM, K-W, CAL [mostly destroyed]), Milde (B), C.B.Clarke (K-W, K, CAL [mostly destroyed], BM), C.Hope (K, K-W [Hope's own herbarium is mainly at BM, P, E, K, DD]), Prantl (B), Christ (P), C.Chr. (BM, K-W, K), Hieron. (B), Lacaita (K-W), Ching (K-W, BM [Ching's own herbarium is in PE, K, BM]). Incidentally it may be noted that though Pichi Sermolli (1996: X, 18) has quoted the present author's name as an example, the present author has never given it himself as Fraser-Jenk., and under principle 14, principle 12 a "may often be broken", as it is in the abbreviation Fras.-Jenk., which the author will retain, rejecting the unwarranted change being put forward as "standard". Brummit (pers. comm. 2003) stated that there was not enough time to correct it, but the present author recommends that henceforth it should be given in the shorter form he standardised.

The later history of the Kew Wallich herbarium (K-W) is well known. While at the Linnean Society, it was reorganised by C.B. Clarke (1832-1908), a school's Inspector in Bengal and Assam, who became a well known expert on Indian ferns and grasses, and later Superintendent of Calcutta Botanic Garden, from 1869-1871. He discovered that quite a number of sheets had become muddled due to having several different collections under one main number, whose subsidiary number was quite often confused (Clarke 1893). This was a serious flaw in the quality of Wallich's documentation work, though much of the problem must have arisen due to careless mounting, which would not have happened had he given a completely different Cat. number for each collection, as strongly pointed out by Clarke. Worse still, if the specimens are small there are often several different collections on the same sheet, with the label only at the top, not indicating which is which sub-number. This problem has led to a number of misunderstandings of Wallichian names by authors unaware of the muddling of a specimen under the same main "Cat. no.", which may have been from one of the other localities under the same number. It has also meant that an author might validate a Wallichian name in a different sense from that originally intended by Wallich. Two examples of this in Pteridophytes are first in *Selaginella*, *Selaginella semicordata* (Wall. *ex* Grev. & Hook.) Spring, which Wallich first named as *Lycopodium semicordatum* from Bechiaco to Dopabasa, banks of the Rapty, [near Hetauda, C. Nepal]. *Wallich* (*Filicologia Nepalensis* 1821:1, *ined*., with full and unequivocal description of erect stem and isomorphic sporophylls *etc*.), *List* no. 126.1, the specimen (sheet 126.3, BM!) being what is now called *S. involvens* (Sw.) Spring. But Greville & Hooker (1833) then validly described the species as with a creeping stem and isomorphic sporophylls, citing both 126.1 [in error] and 126.3, which was evidently the specimen they described. No. 126.3 is listed as a collection of M.R. Smith's from "Montes Sylhet viciniae", which is a quite different, low-altitude, Assamese species, not occurring anywhere near Wallich's Nepalese locality (lectotype (as indicated by Alston 1945): Mountains near Sylhet [*M.R. Smith*] 126/3, E!). To make matters worse, the specimen in K-W labelled 126.3 has been so-labelled in error as it was collected "E Sylhet misit *De Silva* 1821" and is *S. chrysorrhizos* Spring, while a specimen of Greville & Hooker's plant is mixed in with sheet 126.2 in K-W, which is otherwise *S. repanda* (Desv.) Spring! But in fact the type of *S. semicordata* should better be selected from either Hooker's specimens in K (general herbarium) or the material Greville worked most on at E, rather than the material in K-W, which can be thought of as an isolectotype. A second example is *Dryopteris hirtipes* (Bl.) Kuntze subsp.

atrata (Wall. *ex* Kunze) Fras.-Jenk., where Wallich's original material of *Aspidium atratum* was what is now called *D. gamblei* (C.Hope) C.Chr., from woods about Gokurrun, [Gokarna, Kathmandu Valley]. *Wallich* (*Filicologia Nepalensis* 1821, *ined.*), *List* no. 380. But Kunze (1851) first validated the name from the quite different S. Indian material in his possession, formerly at LZ (duplicates often survive at B, K, JE and TUB), applying to the tetraploid, closely related subspecies of *D. hirtipes*.

Clarke also found that there were gaps in the series of numbers, where too much material had been distributed and the Linnean Society therefore printed two request leaflets and sent them out to all the institutions known to have sets of Wallich material, requesting return of specimens in order to fill the gaps (Anon. 1913). Thus it was possible for Clarke to arrange the Linnean set in order and make it almost complete. For the ferns, he also made a list of the Wallichian numbers from N. India (Clarke 1880: 595-606), with their identity, though this was hardly in sufficient detail as it merely gave the single *List* number, name and identity. The list has now been brought up to date in more detail giving the various sub-numbers and their localities (Fraser-Jenkins, in prep.).

The Linnean Society set was given by them to Kew in 1913 and now forms the well known K-W herbarium, which has been microfiched. The Microfiche Company kindly provided a full set of the Pteridophytes to the present author for purchase and he has also studied the original specimens on many occasions; the 507 Pteridophyte numbers of Wallich's *List* are: 1-398, 775, 776, 1031-1037, 2162-2241, 2241, 2680-2685, 4727, 5169, 7073-7091; of these a considerable number are not from India, but from Mauritius and S.E. Asia, and a few are missing or blank sheets (see Clarke 1880: 427-428).

KEY TO RECOGNITION AND SELECTION OF DON'S FERN TYPES

1. Where Don cited a Hamilton specimen:
   a. Original material at BM (Lectotype).
   b. No original material at BM, but in OXF, BR or P-JU *etc.* (Lectotype).
   c. No original material survives (Neotype from other source).

2. Where Don cited a Wallich specimen:
   a. Original early material (1817-1819) surviving at BM (Lectotype).
   b. No original early material at BM, but in OXF, FI-W, P-JU, CGE, B, G, C, LIV, BR (Lectotype)
   c. No original material survives (Neotype from another source).

   (Don often cited both Hamilton and Wallich specimens; either one can be a Lectotype).

4. Where Don cited a Kamroop specimen, search in K-W, K, BM (Lectotype, or if not found, a Neotype from another source).

Picture 1. Hooker's fine illustration of the fern, *Oleandra wallichii* (Hook.) C.Presl, named after his great friend, Wallich. Photographed by the author in 2002; original in the Botany Library, Natural History Museum, London.

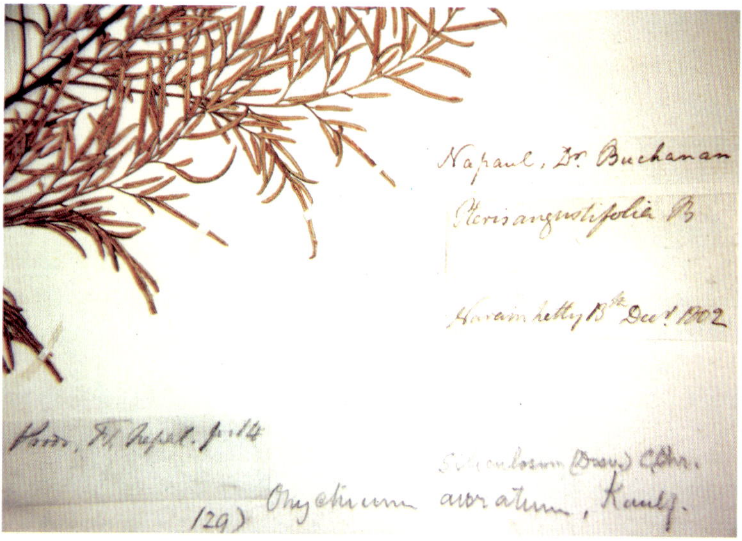

Picture 2. Hamilton's type-specimen in BM of *Onychium siliculosum* (Desv.) C.Chr., with his original label from Narainhetty, 13th Decr. 1802, under the name *Pteris angustifolia* Buchanan, given in his unpublished "*Flora Nepalensis*" of 1802-3. Photographed by the author in 2002.

Picture 3. *Pteris tricolor* Linden, one of the finest of all 19th Century illustrations of ferns, published in 1859 and a true representation of the spectacular colours of young fronds of this magnificent and rare species from S.E. Asia, which just reaches India in Manipur. Photographed by the author in 2002; original in the Botany Library, Natural History Museum, London.

Picture 4. Pteris scabririgens Fras.-Jenk. & T.G. Walker, from E. of Kalimpong, Darjeeling, CRFJ, showing the remarkably pink young fronds. Photographed by the author in May 2004.

Picture 5. Roxburgh's house at Calcutta Botanic Garden, now fallen into disuse, though still structurally sound, where Wallich had his library and herbarium. Photographed by the author in 2004.

Picture 6. The cover of Wallich's lost 1821 manuscript of Nepalese ferns, "*Filicologia Nepalensis*", on its rediscovery by the author in Calcutta herbarium library in 2003. Photographed by the author in 2003.

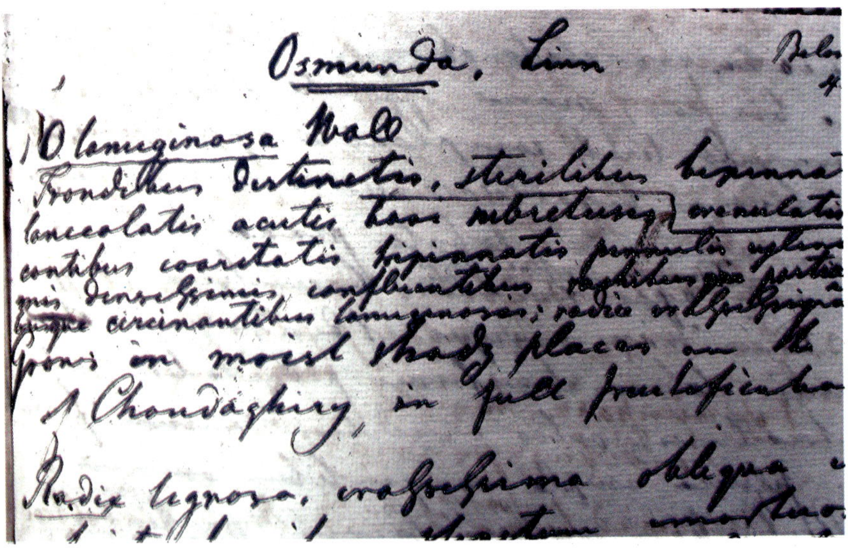

Picture 7. Part of Wallich's own manuscript description of *Botrychium lanuginosum* Wall. ex Hook. & Grev. in his unpublished "*Filicologia Nepalensis*" of 1821, in Calcutta herbarium library. Photographed by the author in 2004.

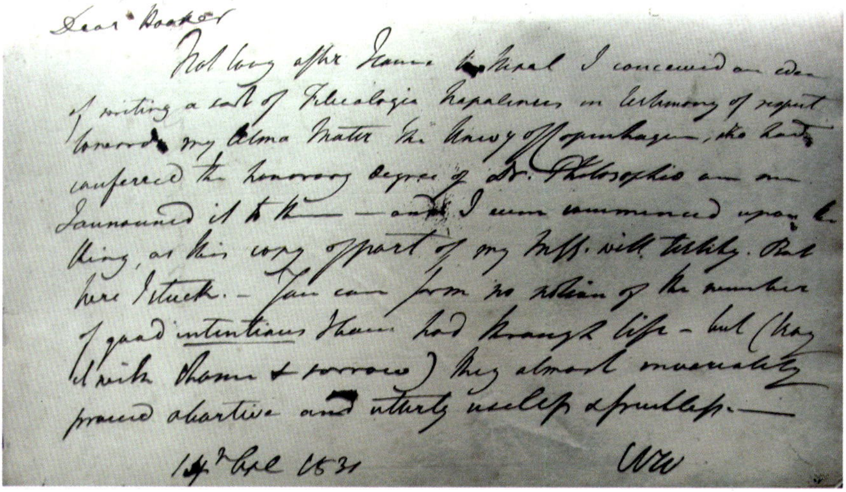

Dear Hooker

Not long after I came to Nepal I conceived an idea of writing a sort of Filicologia Nepalensis in testimony of respect towards my Alma Mater the Univ. of Copenhagen, who had conferred the honorary degree of Dr. Philosophiae on me. I announced it to them — and I even commenced upon the thing, as this copy of part of my MSS. will testify. But here I stuck. — You can form no notion of the number of good intentions I have had through life — but (they [...] with shame & sorrow) they almost invariably proved abortive and utterly useless & fruitless. —

14th Apl 1831 NW

Picture 8. Wallich's apologia on sending his incomplete manuscript to Hooker in 1831, found in the back of "*Filicologia Nepalensis*" in Calcutta herbarium library. Photographed by the author in 2004.

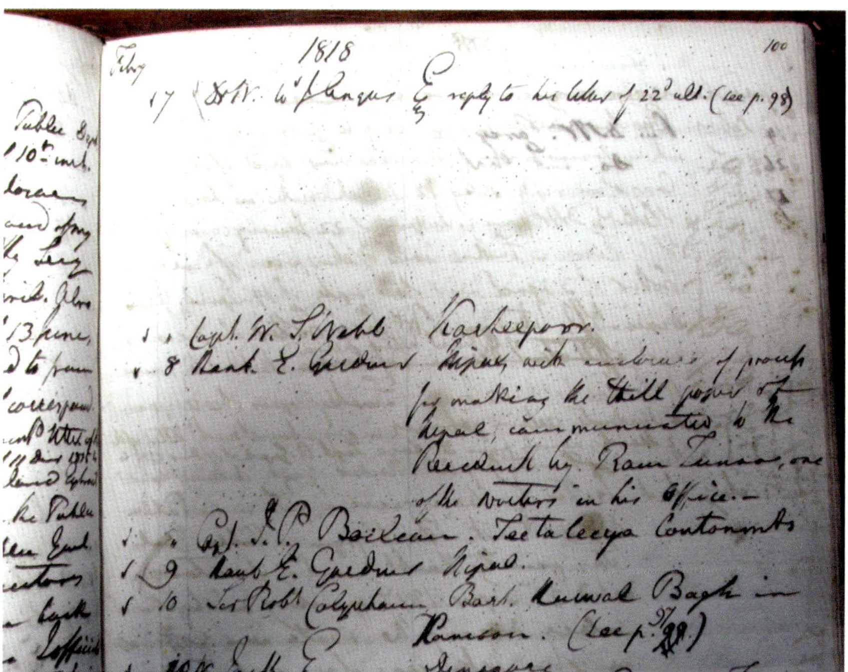

Picture 9. A page of Wallich's handwritten Index-log to his letters at Calcutta herbarium library, listing on 8 Feb. 1818 one of the lost letters from the "Honb. E. Gardner, Nipal", concerning making "*Thill*" or "*til*" spice, used in *achar* or chutney in Nepal. Photographed by the author in 2004.

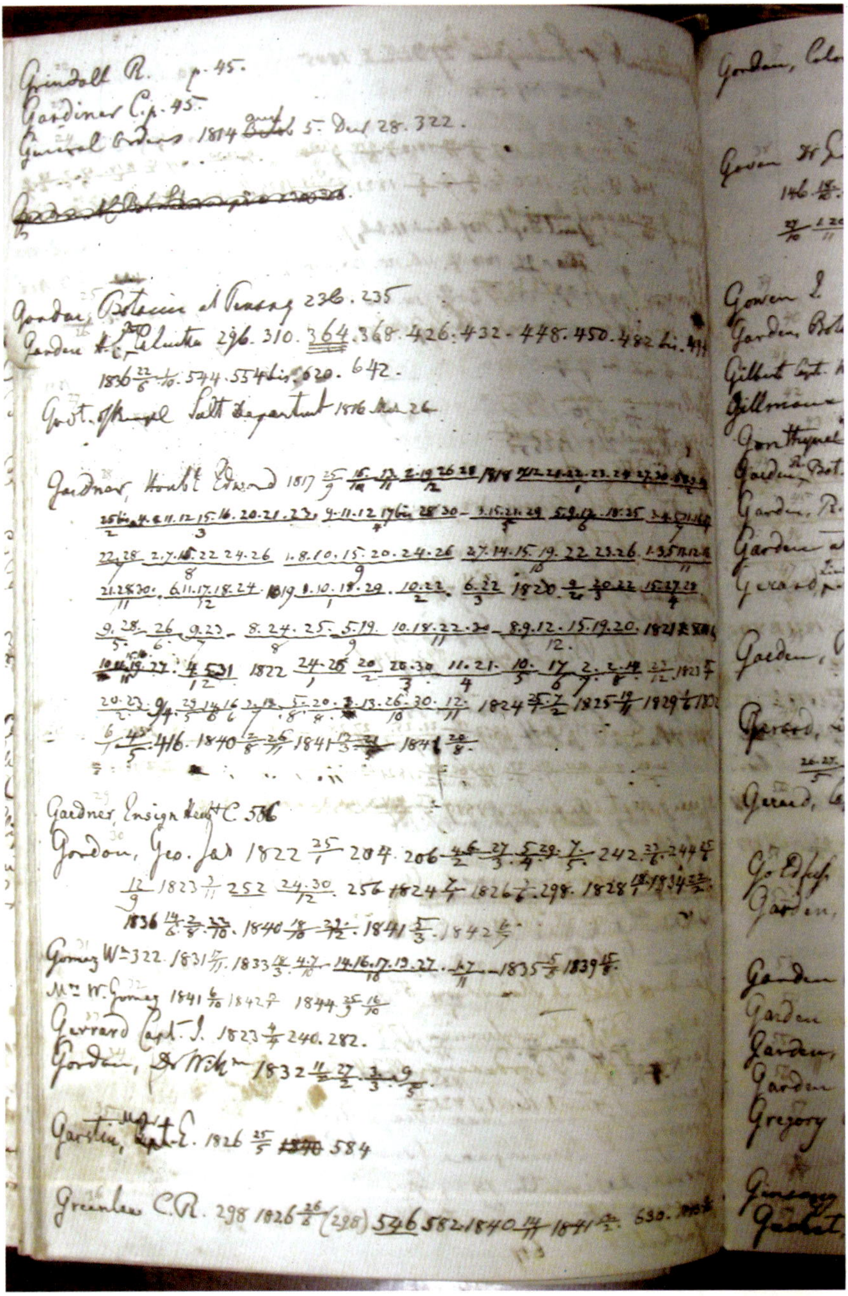

Picture 10. A page from the index to Wallich's letter Index in Calcutta herbarium library, showing some 60 or more lost letters from Gardner in Nepal to Wallich, which gave details of his collections etc. Photographed by the author in 2004.

Picture 11. The back of the spacious, but disused old library built by Sir George King at Calcutta Botanic Garden, and still containing thousands of rare old books, once among them somewhere and perhaps still existing, Wallich's many lost volumes of letters, which would hold the key to his and Gardner's Nepal collections. Photographed by the author in 2004.

Picture 12. The main entrance to the old library at Calcutta Botanic Gardens, still openable by its large 19th Century key. Photographed by the author in 2004.

Picture 13. The south front of the old British Residency building at Lain Khyeo (Lainchaur), adjacent to Narainhetty, in Kathmandu, now replaced by the modern Embassy building. Drawn by Sir Brian Hodgson's brother in *c.* 1833, as it was during Wallich's visit and stay. From Waterhouse (2004), original in Hodgson's scrapbook at the Zoology Library, Natural History Museum, London.

Picture 14. Lt.-Col. William Gardner, the heroic and distinguished first cousin and constant correspondent of the Hon. Edward Gardner. From Dalrymple (2002).

Picture 15. An early map of Kathmandu by Col. Crawford, surveyor with Hamilton on Knox's expedition to Kathmandu in 1802-3, showing in the extreme top-left, the ridge of Nagarjun (Jamachok) mountain, and areas of fields and occasional buildings where the British Residency is situated to the north of the old city. From Phillimore (1950), original in the library of the Survey of India, Dehra Dun.

Picture 16. George Don, brother of Prof. David Don, and himself a well-known botanist in Africa and S. America. From Slack (1990), original in the Library of the Royal Botanic Garden, Edinburgh.

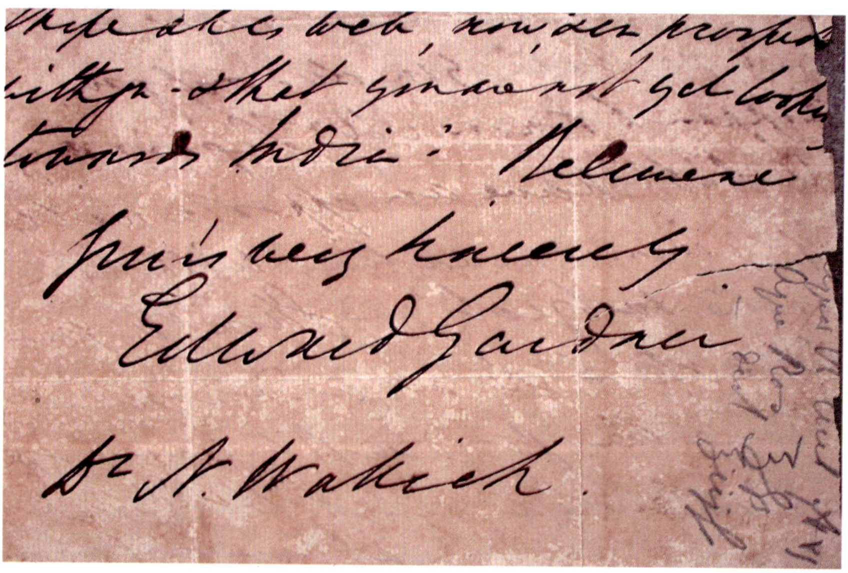

Believe me
yours very sincerely
Edward Gardner

Dr N. Wallich

Picture 17. The Hon. Edward Gardner's signature on a letter to Wallich of 4th May 1832, addressed from his retirement to 20 Bruton Street, London. From one of the four surviving volumes of Wallich's correspondence at Calcutta herbarium library. Photographed by the author in 2004.

Picture 18. Portrait of Dr. Nathaniel Wallich, holding a leaf of the Palm genus, *Wallichia*, named after him by Roxburgh. From the Library of the Royal Botanic Garden, Kew, photographed by Dr. H. Noltie.

Picture 19. *Peranema cyatheoides* D.Don, from Phulchowki Mountain, Kathmandu, CRFJ 30,673, showing its strange hanging Christmas-bauble sori. Don's unfortunate independent naming of this species, discovered and named by Wallich, led to exceptionally strong criticism of him by the normally temperate and kindly Wallich. Photographed by the author in July 2004.

Picture 20. *Asplenium laciniatum* D.Don, from Nagarjun Forest, below Jamachok, Kathmandu, CRFJ 30,603. This is the correct name for the common tetraploid plant around the Kathmandu Valley and the only member of its group there. The name is now generally coming into use instead of its synonym, *A. varians* Hook. Hooker had misunderstood Don's species and description and misapplied his name, creating a nomenclatural problem not resolved until the 1980s. Photographed by the author in July 2004.

Picture 21. *Dryopteris fructuosa* (Christ) C.Chr., from Serbithang, Thimphu, Bhutan, CRFJ 31,485. This fine species with characteristic sport's-jacket button-like indusia was one of the few very distinctive ferns missed by Hamilton, Gardner and Wallich due to their not being allowed to visit higher altitudes in Nepal. Photographed by the author in Oct. 2005.

REFERENCES

Allen, C. 2002. *The Buddha and the Sahibs, The men who discovered India's lost religion*: 1-322. London.

Alston, A.H.G. 1945. An enumeration of the Indian species of Selaginella, *Proc. Nat. Inst. Sci. India* **11** (3): 211-235.

Anon. 1842. The late Professor Don, *Florist's J.* **3**: 15-19.

Anon. 1842. David Don, Esq., *Proc. Linnean Soc. London* **1**: 145-149.

Anon. 1899. William Pamplin (1806-1899), *Proc. Linnean Soc. London* **112** [**110th sess.**]: 80-81.

Anon. 1912. Sir Joseph Dalton Hooker 1817-1911, *Kew Bull. Misc. Info*. **1912**: 1-34.

Anon. 1913. The Wallichian Herbarium, *Bull. Misc. Information Kew* **1913** (7): 255-263.

Arnold, D.A. in press. Nathaniel Wallich.

Bennett, J.J. 1855. Obituary of Nathaniel Wallich, M.D., Ph.D. F.R.S. Lond. & Edinburgh, Vice-President of the Linnean Society, *Proc. Linnean Soc. Lond.* **2**: 314-318.

Blume, C.L. von 1828. *Enumeratio Plantarum Javae et Insularum adjacentium* **1** and **2**. Lugduni Batavorum.

Britten, J. 1902. Buchanan's Avan plants, J. Bot. Brit. For. **40**: 279-282.

Brown, R. 1810. *Prodromus Florae Novae Hollandiae*, London. Reprinted 1960 with Supplementum Primum (1830) and introduction.

Buchanan, F. 1795-1812, *ined. The unpublished Letters of Dr. Francis Buchanan (afterwards Hamilton) addressed to Dr. William Roxburgh, 1795-1812, from the library of Sir David Prain.*

Buchanan, F. 1797, *ined. Enumeratio Plantarum Quas in adeunde civitatem Barmanorum regiam, et dehinc redeundo Anno MDCCXCV observavit Franciscus Buchanan MD.*

Buchanan, F. 1798, *ined. An Account of a Journey undertaken by Order of the Board of Trade through the Provinces of Chittagong and Tiperah, in order to look out for the places most proper for the cultivation of spices, by Francis Buchanan MD.*

Buchanan, F. 1802-1803, *ined. Flora Nepalensis.*

Buchanan, F. *c.* 1803-1804, *ined. Plantarum Nepalensium Icones Pictae.*

Buchanan, F. 1807. *A journey from Madras, through the countries of Mysore, Canara and Malabar, performed under the orders of the Most Noble the Marquis of Wellesley, for the express purpose of investigating the state of Agriculture, Arts and Commerce, the Religion, Manners and Customs, the Natural and Civil History, and Antiquities in the Dominions of the Rajah of Mysore, and the countries acquired by the Honourable East India Company in the late and former wars, from Tippoo Sultan* **1**, **2** and **3**. London.

Buchanan, F. *c.* 1815, *ined. Catalogue of dried plants collected, presented to the Museum of the Hnble. East India Company, and arranged according to the system of Linnaeus, By Francis Buchanan MD.*

Burke, J. & Sir B. 1967. *Burke's Genealogical and Heraldic History of the Peerage, Baronetage and Knightage* (ed. 104, P. Townend): Gardner (Dormant), 1012-1014. London.

Burkill, I.H. 1953-1962. Chapters on the history of botany in India, 1-4, *J. Bombay Nat. Hist. Soc.* **51** (4): 846-878 (1953); **54** (1): 42-86 (1956); **58** (3): 678-796 (1960); **59** (2): 335-359 (1962).

de Candolle, R. 1981, *ined. Nathaniel Wallich. How the largest Botanical collection yet brought to Europe by a single man was distributed.*

de Candolle, R. & Radcliffe-Smith, A. 1981. Nathaniel Wallich, MD, PhD, FRS, FLS, FRGS, (1786-1854) and the Herbarium of the Honourable East India Company, and their relation to the de Candolles of Geneva and the *Great Prodromus*, *Bot. J. Linnean Soc.* **83**: 325-348.

Chatterjee, D. 1948. National Herbarium for India, *Sci. & Culture* **13** (8): 328-330.

Ching, R.-C. 1984. Problems in Indian Ferns, *Indian Fern J.* **1**: 47-50.

Christensen, C. 1881. Nathaniel Wallich (forhen Nathan Wollf) [Bibliography], *Bot. Tidsskr.* **12**: 105-106.

Christensen, C. 1924, 1925, 1926. [Life of] Nathaniel Wallich, in *Den Dansk Botaniks Historie med tilh. Bibliografi etc.* **1**: 210, 251-253, 354, 501; **2**: 157-160.

Clarke, C.B. 1880. A Review of the Ferns of Northern India: with an index of the species and 36 plates, *Trans. Linnean Soc. London*, ser. 2 Bot., **1**: 425-619, *tt.* 49-84. Reprinted 1973, Dehra Dun and Delhi

Clarke, C.B. 1893. Collector's numbers, *J. Bot. Brit. For.* **31**: 135-138.

Clokie, H.N. 1964. *An account of the herbaria of the Department of Botany in the University of Oxford*: 1-228 et tt. Oxford.

Dalrymple, W. 2002. *White Mughals, Love and Betrayal in Eighteenth-Century India*: 1-580 *et tt.* London.

Dalrymple, W. 2003. Peer into the past, *The Guardian* [London newspaper] 8 Dec. 2003.

Dawson, W.R. 1934. Sir J.E. Smith Papers. London

Debrett, J. 1825. *Debrett's Peerage and Baronetage*: *sub* Blessington.

Desmond, R.G.C. 1966. John Smith, Kew's first curator, *J. Kew Guild* **8**: 576-587.

Desmond, R. [G.C.] 1992. *The European Discovery of the Indian Flora*. Kew and Oxford.

Desmond, R. [G.C.] 1994. *Dictionary of British & Irish Botanists and Horticulturalists Including Plant Collectors, Flower Painters and Garden Designers*, ed. 2: 1-825. London and Bristol.

Don, D. 1820. Descriptions of several new plants from the Kingdom of Nepaul: taken from specimens preserved in the herbarium of Aylmer Bourke Lambert, Esq. communicated by Mr. David Don, *Mem. Wernerian Nat. Hist. Soc. Edinburgh* **3**: 407-415, *et tt.*

Don, D. 1824. *Prodromus Floræ Nepalensis, sive Enumeratio Vegetabilium quæ in itinere per Nepaliam proprie dictam et regiones conterminas, ann. 1802-1803, detexit atque legit D. D. Franciscus Hamilton (olim Buchanan) M.D. Societ. Reg. et Linnæan. Londin. Soc. Accedunt plantæ a D. Wallich nuperius missæ, secundum methodi naturalis normam disposuit atque descripsit.* Londinium.

Don, D. 1825 ["1823"]. Descriptions of Nine new Species of the Genus Carex, Natives of the Himalayan Alps in Upper Nepal, *Trans. Linnean Soc. London* **14**: 325-333.

Don, D. 1828. An account of the Lambertian herbarium by Mr. Don, in Lambert, A.B., *A Description of the Genus Pinus, illustrated with figures; directions relative to the cultivation, and remarks on the uses of the several species, also descriptions of many other trees of the family of Coniferæ ...... to which is added an appendix containing descriptions and figures of some other remarkable plants, and an account of the Lambertian herbarium, by Mr. David Don, Libr. L.S. etc.*, ed. 2, **2**: App. 13-24. London.

Don, D. 1841. A monograph of the genus Disporum, *Trans. Linn. Soc. Lond.* **18**: 513-524.

Don, W.G. 1897. *Memoirs of the Don family in Angus: with a general survey of the etymology of the name, and of the Scottish family; also with archæological appendices*. London.

Fraser-Jenkins, C.R. 1977. *New Species Syndrome in Indian Pteridology and the ferns of Nepal*: 1-403 & Erratae. Dehra Dun.

Fraser-Jenkins, C.R. 2004. The rediscovery of two lost 19th Century fern books by Moore and Wallich, *Natural Sci. Collections Assoc. News* **3**: 47-51.

Fraser-Jenkins, C.R. in prep. *Wallich's unpublished Filicologia Nepalensis vel Tentamen de filicibus Napalensibus, sistens descriptiones et icones filicum in itinere Napalensi observatorum adjectis speciebus Indiae orientalis et insulae Mauritii minus cognitis vel novis* (*1821*).

Fraser-Jenkins, C.R., in prep. *The Pteridophytes of David Don's Prodromus Floræ Nepalensis* (*1824*).

Gage, Major A.T. & Burkill, I.H. 1916. William Jack's Letter's to Nathaniel Wallich, 1819-1821, *J. Straits Branch Roy. Asiatic Soc.* **73**: 147-242.

Gardner, W.L. 1820-1821, *ined.* The Gardner Papers (letters of Col. William Linnaeus Gardner, Khassgunge, to his cousin the Hon. Edward Gardner, Katmandu, with notes by William Dalrymple), typed transcription of originals in Indian Office Library. National Army Museum ref. 6305-6356.

Ghosh, A.K. 1947-1948. National Herbarium for India, *Sci. & Culture* **13** (1): 75-77 (1947); **13** (8): 330-331 (1948).

Gilmour, J.S.L. & Tutin, T.G. 1933. *A list of the more important collections in the University Herbarium, Cambridge.*

Gould, T. 1999. *Imperial Warriors, Britain and the Gurkhas*: 1-480. London.

Gray, J.L. 1893. *The letters of Asa Gray.* Boston, Mass.

Greville, R.K. & Hooker, W.J. 1833. Enumeratio Filicum, Part 1, Lycopodineae, *Hook. Bot. Miscellany* **2**: 360-403, **3**: 216-250.

Griffith, W. 1843, *ined. Report on the Botanic Garden, Calcutta.*

Griffith, W. 1844. The Cryptogamous Plants of Dr. ROXBURGH, forming the fourth and last part of the Flora Indica, *Calcutta J. Nat. Hist.* **4**: 463-520 and index xv-xvi.

Hamilton, F. 1819. *An account of the Kingdom of Nepal, and of the territories annexed to its dominion by the House of Gorkha*: 1-316. Edinburgh. Reprinted 1971. New Delhi.

Hamilton, F. 1820. An Account of Asam with some notices concerning the neighbouring territories, *Annals Oriental Literature*; first compiled in 1807-1814 by Francis Buchanan, M.D., F.R.S; **1820**: 244- . Republished 1940, ed. S.K. Bhuyan. Gauhati.

Hamilton, F. 1821. Autobiographical notes, *Trans. Roy. Soc. Edinburgh* **10**: 171 *et seq.*

Hamilton, F. 1822. *An Account of the fishes found in the river Ganges and its branches* **1** & **2**. Edinburgh.

Hamilton, F. 1822-1837. A commentary on the Hortus Malabaricus. 1-4. *Trans. Linnean Soc. London* **13** (2): 474-560 (1822), **14** (2): 171-312 (1825), **15** (1): 78-152 (1827), **17**: 147-252 (1837).

Hamilton, F. 1824-1829. Commentary on the Herbarium Amboinense, 1-2. *Mem. Wernerian Soc. Edinburgh* **5**: 307-383 (1824), **6**: 268-333 (1829).

Hamilton, F. 1826. Some notices concerning the Plants of various parts of India, and concerning the Sanscrita names of those Regions, *Trans. Roy. Soc. Edinburgh* **10**: 171- .

Hara, H. 1985. On typification of David Don's names, I.E., *Cymbidium longifolium* (Orchidaceae), *Taxon* **34**: 690-691.

Hara, H., Stearn, W.T. & Williams, L.H.J. 1978. *An Enumeration of the Flowering Plants of Nepal* **1**: 1-22 *etc.* (Introduction by W.T. Stearn; Plant collectors in Nepal by S. Sutton).

Hasrat, B.J. 1970. *History of Nepal As Told by its Own and Contemporary Chroniclers*. V.V. Research Institute, Hoshiarpur, Punjab.

Hill, Sir A.W. 1939. The Indian collections at Kew, and the relations between Kew and Sibpur, *Proc. Indian Sci. Congr.* **25** (4): 133-135.

Holttum, R.E. 1967. John Smith of Kew, *Brit. Fern Gaz.* **9** (8): 330-334 *et t.*

Hooker, J.D. & Thomson, T. 1855. Introductory essay, in *Flora Indica*: 1-280. London.

Hooker, W.J. 1820. *Musci Exoticae; containing figures and descriptions of new or little known foreign mosses and other cryptogamic subjects* **2**: *t.* 146. London.

Hooker, W.J. 1842. Sale of Mr. Lambert's Herbarium, *London J. Bot.* **1**: 394-396.

Hooker, Sir W.J. 1854. Dr. Wallich, *Hook. J. Bot. & Kew Gard. Misc.* **6**: 185.

Hooker, W.J. 1855. Botanical information, Dr. Wallich, Hook. *J. Bot. & Kew Gard. Misc.* **6**: 185.

Hunter, Sir W.W. 1896. *Life of Brian Houghton Hodgson, British Resident at the Court of Nepal etc.*: 1-390. London.

International Code of Botanical Nomenclature. 2000. Greuter, W., McNeill, J., Barrie, F.R., Burdet, H.M., Demoulin, V. Filgueiras, T.S., Nicolson, D.H., Silva, P.C., Skog, J.E. Trehane, P., Turland, N.J. & Hawksworth, D.L. (eds.) International Code of Botanical Nomenclature (Saint Louis Code) adopted by the Sixteenth International Botanical Congress St. Louis, Missouri, July-August 1999, *Regnum Vegetabile* **138**: 1-474. Königstein.

Jack, G.B. 1835, William Jack, *Hook. Comp. Bot. Mag.* **1**: 120-140.

Jack, W. 1822. *Malayan Miscellanies* **2** (7): 1-96.

Jackson, V.H. 1925. *Journal of Francis Buchanan (afterwards Hamilton) kept during the Survey of the Districts of Patna and Gaya in 1811-1812.* Patna.

Kew Director's Correspondence **52**: *f.* 47 (13 Oct. 1818).

Khanduri, C.B. 1997. *A re-discovered history of Gorkhas*: 1-306. Delhi.

Khullar, S.P. 2000. *An Illustrated Fern Flora of the West Himalaya (with some contributions and assistance from Mr. C.R. Fraser-Jenkins)* **2**: 542-544 (Appendix 2). Dehra Dun.

King, Sir G. 1899. The early history of Indian Botany, *J. Bot. Brit. For.* **37**: 458-459; A Sketch of the History of Indian Botany, *Gard. Chron.*, ser. 3, **26**; *Rep. Brit. Assocn.Adv. Sci., Dover* **1899**: 904-919.

Kirkpatrick, Col. [W.] 1811. *An account of the Kingdom of Nepaul, being the substance of observations made during a Mission to that country in the year 1793*: 1-388. London. Reprinted 1996, New Delhi and Madras.

Kunze, G. 1851. Filices Nilagiricae, *Linnaea* **24**: 239-299.

Laird, M. 1988. *Thomas Moore Curator at Chelsea Physic Garden 1848-1887*. London and Omagh.

Lambert, A.B. 1802-1842. *Description of the Genus Pinus etc.*, eds. 1-5, **1-3**. London.

Landon, P. 1928. *Nepal* **1**: 1-358; **2**: 1-363. London. Reprinted 1993, Asian Educational Services, New Delhi.

Lanjouw, J. & Stafleu, F.A. 1957. Index Herbariorum, part 2 (2) Collectors, E-H, *Regnum Vegetabile* **9**: 197.

Lee, S. (ed.) 1892. Aylmer Bourke Lambert (1761-1842), *Dictionary of National Biography* **32**: 6-7. London.

Lee, S. (ed.) 1899. Nathaniel Wallich (1786-1854), *Dictionary of National Biography* **59**: 135-136. London.

Lindley, J. 1825. Note, *Bot. Register* **11**: *sub t.* 872.

Lindley, J. 1842. Mr. Lambert's Sale. Sale of Mr. Lambert's Herbarium, *Gard. Chronicle Agric. Gaz* **1842**: 17, 27.

Luxford, G. 1844. Death of Professor Don [with bibliography], *The Phytologist* **1**: 133-134

Mabberley, D.J. 1977. Francis Hamilton's Commentaries with particular reference to Meliaceae, *Taxon* **26** (5-6): 523-540.

Mabberley, D.J. 1980. A re-examination of the 'Indian Catalogues', in K.S. Manilal (ed.), *Botany and history of Hortus Malabaricus*: 80-110. Oxford.

Madden, Lt.-Col. E. 1856/8. Elucidation of some plants mentioned in Dr. Francis Hamilton's account of the Kingdom of Nepal, *Trans. Bot. Soc. Edinburgh* **5**: 116-140.

Martin, M. 1838. *The History, Antiquities, Topography and Statistics of Eastern India, comprising the districts of Behar, Shahabad, Bhagalpoor, Goruckpoor, Dinagepoor, Puraniya, Rungpoor, and Assam; in relation to their Geology, Mineralogy, Botany, Agriculture, Commerce, Manufactures, Fine Arts, Population, Religion, Education, Statistics, &c., surveyed under the orders of the Supreme Government, and collated from the original documents at the East India House with the permission of the Honourable Court of Directors, by Montgomery Martin* **1**, **2** and **3**. London.

M'Clelland [McClelland], J. 1841. [re] Report on the Botanic Garden of Calcutta by Dr. Wallich, Supt., *Calcutta J. Nat. Hist.* **2** (6): 288-294.

M'Clelland [McClelland], J. 1848. *Memorandum regarding the differences between Dr. Wallich & the late W. Griffith.* Calcutta.

Miller, H.S. 1970. The herbarium of Aylmer Bourke Lambert notes on its acquisition, dispersal, and present whereabouts, *Taxon* **19** (4): 489-553.

Moore, T. 1857-1862. *Index Filicum, being a Synopsis of the Genera of Ferns with Characters and Illustrations and an Enumeration of the Species with geographical Distribution, Synonyms, References to Figures, Descriptions etc.* **1** and **2**. London.

Murray, G., Britten, J. & Gepp, A. 1904. The Department of Botany, in Lankester, E.R., *The History of the collections contained in the Natural History Departments of the British Museum* **1**: 1-193. London.

Nayar, M.P. & Das, A.R. 1983. William Roxburgh (1751-1815) & Nathaniel Wallich (1786-1854), *Bull. Bot. Surv. India* **25**(1-4): 347-356.

Noltie, H.J. 1999. *Indian Botanical Drawings 1793-1868 from the Royal Botanic Garden Edinburgh*: 1-100. Edinburgh.

Nooteboom, H. 1997. The microsoroid ferns (Polypodiaceae), *Blumea* **42**: 261-395.

Nugent, Lady M. 1839. *Journal of a Residence in India 1811-1815* **2**: 9. London.

Parkes, F. 1850. *Wanderings of a Pilgrim in Search of the Picturesque*. London. Reprinted 2003, London.

Phillimore, Col. R.H. 1945-1959. *Historical Records of the Survey of India* **1** 18th Century (1945); **2** 1800 to 1815 (1950); **3** 1815-1830 (1954); **4** 1830-1843, George Everest (1958); Waugh, A. **5** 1844-1861 (1959).

Pichi Sermolli, R.E.G. 1992. Pteridophytes, in Brummitt, R.K. & Powell, *Authors of Plant Names*. Kew.

Pichi Sermolli, R.E.G. 1996. *Authors of scientific names in Pteridophyta*: 1-78. Kew.

Prain, D. 1905. A Sketch of the Life of Francis Hamilton (once Buchanan), some time Superintendent of the Honourable Company's Botanic Garden at Calcutta, *Annals Roy. Bot. Gard. Calcutta* **10** (2): i-lxxv.

Press, J.R. & Shrestha, K.R. 2000. Collections of flowering plants by Francis Buchanan-Hamilton from Nepal, 1802-1803, *Bull. Nat. Hist. Mus. London* (Bot.) **30** (2): 101-130.

van Reede [Rheede] tot Draakestein [Drakenstein], H.[A.] 1678-1693 ["1703"]. *Hortus Indicus Malabaricus* **1-12**. Amstelodami. [Vol. **12** contains the Pteridophytes].

Renkema, H.W. & Ardagh, J. 1930. Aylmer Bourke Lambert and his 'Description of the Genus Pinus', *J. Linnean Soc. London* **18**: 439-466 (with appendix by S. Savage).

Roxburgh, W. 1820-1824. *Flora Indica*, ed. 1, **1** and **2**. Serampore.

Rumphius [Rumph], G.E. 1741-1755. *Herbarium Amboinense* (*Het Amboinesch Kruid-boek*), *Plurimus conplectens Arbores, Frutices, Herbas, Plantas terrestres & aquaticas, quae in Amboina, et adjacentibus reperiuntur insulis, Adcuratissime descriptas juxta earum formas, cum diversis denominationibus, cultura, usu, ac virtutibus. etc. Cura & Studio Joannis Burmanni.* **1-7**. Amstelædami. [Vol. **6** contains the Pteridophytes].

Savage, S. 1963, *ined*. Catalogue of the Smith Herbarium **6**. Pteridophyta.

van Schendel, W. 1992. *Francis Buchanan in southeast Bengal* (*1798*) *His Journey to Chittagong, the Chittagong Hill Tracts, Noakhali and Comilla*: 1-209. University Press, Dhaka.

Schultes, J.A. 1830. Schultes's botanical visit to England, *Hook. Bot. Misc.* **1**: 48-78 (with introduction by W.J. Hooker).

Sen, Dr. J. 1962. The Indian Botanic Garden, Sibpur, Calcutta, *Taxon* **11** (7): 212-213.

Shrestha, K.K. & Press, J.R. 2000. *Catalogue of Type Specimens from Nepal.* London.

Slack, A. 1990. George Don of Forfar (1764-1814), *The Rock Garden, J. Scottish Rock Gard. Club* **22**: 44-50.

Smith, A.R[eid]. & Fraser-Jenkins, C.R. 1982. Dryopteris paleacea is a synonym of D. wallichiana, *Taxon* **31** (2): 326-329.

Smith, A.R. & Parsons, T.J. 1970. *Prefatory notes for the microfiche of the herbarium of the East India Company ('The Wallich Herbarium') at Kew*. Zug.

Smith, J.E. 1804-1808. *Exotic Botany, consisting of coloured Pictures and scientific Descriptions of such new, beautiful or rare plants as are worthy of Cultivation* **1**: *tt.* 1-60 (1804-1805), **2**: *tt.* 61-120 (1805-1808). London.

Smith, J.E. 1804, 1808. [Descriptions of Flowering Plants] in Rees, A., *The Cyclopædia; or, Universal Dictionary of Arts, Sciences, and Literature* **3**, **11**. London.

Smith, Lady P. 1832. *Memoir and Correspondence of the late Sir James Edward Smith* **1**: 555, **2**: 85-88. London.

Sotheby, S.L. 1842. *Catalogue of the highly valuable Botanical Museum of the late A.B. Lambert, Esq. of Boyton House, Wiltshire, &c. ..... Which, by order of the Executor, will be sold by Auction, at 26, Lower Grosvenor Street (the residence of the late Mr. Lambert) By Mr. S. Leigh Sotheby, Auctioneer of Literary Property, Wellington Street, Strand, on Monday, June 27th, 1842, and Two following Days, at One o'Clock, precisely. To be viewed six days previous to the Sale, from Ten o'Clock in the morning till Five in the Evening*: 1-16. London.

Stafleu, F.A. 1972. HI-IAPT Portraits of Botanists, no. 27. Aylmer Bourke Lambert, *Taxon* **21** (5/6): 614.

Stafleu, F.A. & Cowan, R.S. 1976-1988. *Index to Taxonomic Literature*, ed. 2, **1**: 668-669 (1976); **2**: 35, 735-737 (1979); **3**: 571-573 (1981); **4**: 40-41 (1983); **7**: 37-41 (1988). Utrecht and The Hague.

Stafleu, F.A. & Mennega, E.A. 2000. *Taxonomic Literature Supplement* **6**: 48-49.

Stearn, W.T. 1945. David Don's "Prodromus Florae Nepalensis", *J. Arnold Arboretum* **26**: 168.

Stearn, W.T. 1960. Allium and Milula in the Central and Eastern Himalaya, *Bull. Brit. Mus. Nat. Hist.*, Bot. **2**: 159-191.

Stearn, W.T. 1978. Introduction, in Hara, H., Stearn, W.T. & Williams, J.H., *Enumeration of the Flowering Plants of Nepal* **1**: 1-22.

Stearn, W.T. 1988. James Edward Smith (1759-1828): first President of the Linnean Society and his herbarium, *Bot. J. Linnean Soc.* **96**: 199-216.

van Steenis-Kruseman, M.J. 1958. Malaysian Plant Collectors and Plant Collections Supplement 1, *Flora Malesiana*, ser. 1, **5** (4).

van Steenis-Kruseman, M.J. & van Steenis, C.G.G.J. 1950. Malaysian Plant Collectors and Plant Collections, Suppl. 1, *Flora Malesiana*, ser. 1, **5** (4): ix-cxlii, 1-120.

Steinberg, C.H. 1977. The collectors and collections in the Herbarium Webb, *Webbia* **32** (1): 1-49.

Stephen, L. (ed.) 1886. Francis Hamilton Buchanan, MD (1762-1829), *Dictionary of National Biography* **7**: 186. London.

Stephen, L. (ed.) 1888. David Don (1800-1841), *Dictionary of National Biography* **15**: 204-205. London.

Stephen, L. (ed.) 1890. William Linnaeus Gardner (1770-1835), *Dictionary of National Biography* **20**: London.

Sterll, M. in prep. Nathaniel Wallich.

Stiller, L.F. 1973. *The Rise of the House of Gorkha: a study in unification of Nepal 1768-1816*. New Delhi.

Symes, M. 1800. *An Account of an Embassy to the Kingdom of Ava*. London.

Taylor, R. 1842. Obituary: Professor Don, *Annals Mag. Nat. Hist.* **8** (52): 397-399.

Taylor, R. 1842. Some Notices of the Late Professor Don, and of his Father, Mr. George Don, formerly Curator of the Edinburgh Botanic Garden, *Annals Mag. Nat. Hist.* **8** (53): 476-479.

Thomson, T. 1857. Notes on the herbarium of the Calcutta Botanic Garden, with especial reference to the completion of the Flora Indica, *J. Roy. Asiatic. Soc. Bengal* **25**: 405-418.

Thiselton-Dyer, Sir W.T. 1891. Historical account of Kew to 1841, *Kew Bull. Misc. Inf.* **1891**: 270-327.

Underwood, L.M. 1905. The early writers on ferns and their collections.- 4. Presl, 1794-1852; John Smith, 1798-1888; Fée, 1789-1874; and Moore, 1821-1887, *Torreya* **5** (3): 37-41.

Vegter, I.H. 1986. Index Herbariorum, part 2 (6) Collectors, S, *Regnum Vegetabile* **114**: 938.

Wallich, N. 1818. Descriptions of some rare Indian plants, and An Account of a new species of Camellia growing wild at Nepal, *J. Asiat. Res.* **13** (4): 369-445 *etc.*

Wallich, N. 1821, *ined. Filicologia Nepalensis, Tentamen de filicibus Napalensibus, sistens descriptiones et icones filicum in itinere Napalensi observatorum adjectis speciebus Indiae orientalis et insulae Mauritii minus cognitis vel novis.*

Wallich, N. 1823. Aus einem Brief von Dr. Wallich, Direktor des botanischen Gartens in Calcutta an Prof. J.W. Hornemann, dat. Katmandu in Napal, d. 18 Juli. 1821, *Flora* **6** (1): 65-75.

[Wallich, N.] 1854. [Dr. Wallich Obituary], *Gard. Chronicle* **1854**: 284.

Wallich, N. 1824-1826. *Tentamen floræ napalensis illustratæ* **1** and **2**. Calcutta and Serampore.

Wallich, N. 1828-1849. *A numerical List of dried specimens of plants in the East India Company's Museum, collected under the superintendence of Dr. Wallich of the Company's Botanic Garden at Calcutta*: 1-253 (nos. 1-9148). London and Hereford (lithographed by N. Wallich and G. Bentham).

Wallich, N. 1829, 1830, 1832. *Plantae Asiaticae Rariores* **1**: 33, 35, 38, 42; **2**. London. Reprinted 1979. Dehra Dun.

Waterhouse, D.M. (ed.) 2004. *The origins of Himalayan studies, Brian Houghton Hodgson in Nepal and Darjeeling 1820-1858*: 1-280. London & New York.

Whelpton, J. 1991. *Kings, Soldiers and Priests, Nepalese Politics 1830-1857*: 1-315. New Delhi.

Wight, R. 1831. Illustrations of Indian Botany, principally of the southern parts of the peninsula, *Hook. Bot. Miscellany* **2**: 90-97.

In Dec. 2005, the author was delighted to find at last the missing earlier Wallich correspondence in the new library at Calcutta Botanical Garden, with the kind help of Dr. B. Prasanna. The 17 additional complete volumes in good condition run from 1794 to 1830 and contain all details of the early and later Nepal collections, with large numbers of letters concerning Nepal from the Hon. Edward Gardner, Robert Stuart, the Hon. Gerald Wellesley, Lt. Peter Boileau, Francis De Silva, Sir Robert Colquhoun, Sir William Hooker, Dr. Francis Hamilton, Sir James Smith, Prof. David Don, Lt. William Jack, Sir Brian Hodgson, Prof. W. Hornemann and others. These are now being studied in detail and transcribed by the author with a view to further publication of early Nepalese botanical discovery from the remarkable Botanical Embassy to Nepal.

The whole collection of Wallich letters and papers are currently being photographed in Calcutta by the author's colleagues, Dr. Michael Sterll and Prof. Ib Friis, of Copenhagen University and Botanical Museum, for preservation and further study in connection with their forthcoming definitive biography of Wallich.

INDEX

Page-numbers in bold are for the major botanists covered in this work.